The Book of
RHODODENDRONS

The Book of
RHODODENDRONS

Marianna Kneller

TIMBER PRESS
Portland, Oregon

To Stephen

KEY TO FLOWER PAINTINGS

a flowering branch

b flower (corolla)
  b*i* flower to show scales
  b*ii* flower base

c petal (corolla lobe)
  c*i* petals
  cii petals undersurface
  c*iii* petals & calyx

d section through flower

e stamen
  e*i* stamens
  e*ii* stamen enlarged
  e*iii* stamens, pistil & calyx
  e*iv* stamens & pistil

f calyx
  f*i* calyx cross-section
  f*ii* calyx & pistil
  f*iii* calyx with scales

g capsule
  g*i* capsules
  g*ii* capsule cross-section
  g*iii* capsule cross-section with seeds

g*iv* capsule husk
g*v* capsule & calyx
g*vi* capsule & pistil
g*vii* capsule enlarged to show scales
g*viii* capsule enlarged

h ovary
  h*i* ovary cross-section
  h*ii* ovary enlarged
  h*iii* pistil (ovary, stigma & style)
  h*iv* pistil enlarged
  h*v* pistil & calyx
  h*vi* pistil & calyx enlarged to show scales

j flower bud
  j*i* flower bud scales
  j*ii* flower with bud scales

k leaf
  k*i* leaf bud
  k*ii* leaf cross-section
  k*iii* leaf undersurface
  k*iv* leaf, upper to show hairs
  k*v* leaf, part to show scales
  k*vi* leaf scales
  k*vii* leaf scale(s) enlarged
  k*viii* leaf base

k*ix* recurve of leaf
k*x* apex of leaf, undersurface to show scales
k*xi* leaf undersurface to show scales

l new leaf
  l*i* new leaf growth

m stigma
  m*i* stigma & style

n hair

o anther

  o*i* anthers

p petiole cross-section

q winter bud

r inflorescence

s autumn foliage

t pedicel

Editorial note regarding botanical classifification and nomenclature
The Edinburgh Botanic Garden *Revision of Rhododendron* (1980 onwards) is used in this book wherever possible.
However, both naming and subsequent necessary changes, including re-classification, have been going
on for so long now that some earlier names and groupings have become widely accepted, particularly among practising gardeners.
Therefore, small inconsistencies may occasionally occur.

Designed by Lisa Tai

Typeset by ABM Typographics Limited

Printed in Singapore by CS Graphics PTE Limited

First published in North America in 1995 by
Timber Press. Inc.
The Haseltine Building
133 S.W. Second Avenue, Suite 450
Portland, Oregon 97204, U.S.A.

ISBN 0-88192-322-2

Page 1: Flower bud of *R. crassum*; page 2: *R. ponticum* Linn. "from Pontus, Asia Minor";
page 7: *R. crassum* Franch. "fleshy", yīn mai dùjuān, "hidden-vein rhododendron"

CONTENTS

FOREWORD

*I*n the long and distinguished history of horticultural literature, books combining informative text with reproductions from paintings – whether for botanical or for aesthetic reasons – are particularly esteemed. The eye of the artist supplies a personal element which is transferred from the flat page to that human mechanism which enables us to appreciate beauty to our own satisfaction. This personal view is immediately clear in Marianna Kneller's paintings of her beloved rhododendrons. However, in this book she has added another dimension, an instructive element. The genus is an enormous one, so almost inevitably it is also complex.

The rhododendron taxonomists have studied, described and classified many thousands of wild-collected specimens, and by publication of their conclusions have made them available, not only to other botanists but to the general public. But with several hundred species and their variants, and suspected natural hybrids, illustration on the scale required is impractical. However, their methods of dividing, grouping and classifying the species simplify the process of reaching a framework understanding of the distinctions involved. Thus, Marianna Kneller has used the methods of the taxonomists of the Royal Botanic Garden, Edinburgh, to shed light on the wide range of species illustrated in this selection of her paintings.

The living material she needed for her paintings came mainly from the gardens at Exbury, but some was obtained from other rhododendron growers. Some of these growers were previously unknown to her, but when she proposed that they write their own specialist articles for the text, revealing both their experience and their enthusiasm, they responded willingly. The outcome is this splendidly original book, surely unique in the literature of the rhododendron.

KENNETH LOWES

a

b

hiii

j

k

d

ei

ci

g

PART I
SUBGENUS HYMENANTHES

*T*he 200-acre garden and land of Bassett Wood House was my playground. It might sound very grand but, in truth, the gardens were neglected by the Second World War and its consequences.

In springtime, I wandered among primroses, violets and wood anemones which grew in profusion on the banks of the streams that fed the Top and Lower Lily Ponds. A few weeks later, delicate forget-me-nots would fringe these same streams and huge deep-golden marsh marigolds would welcome the early summer sun. Carpets of sweet-smelling bluebells covered the ground under the silver birch and hazel coppices. Among the wild cherry trees, budding brambles, willowy catkins and white furry pussy-willow, buds would soon emerge, but the finest prize of all was the huge, exquisitely coloured wild rhododendron, *R. ponticum*.

Seventeen miles away, on the eastern side of the Beaulieu River, another Estate was slowly recovering from its wartime occupation. The lovely House and Gardens of Exbury, the home and creation of Lionel de Rothschild, had been requisitioned by the Admiralty in 1942 and played a significant part in the D-Day Landing. Lionel de Rothschild was an exceptional man. He was a Founder Member of the Rhododendron Association, its first President and Editor of its *Year Book*. Among all the plants he collected and grew at Exbury, rhododendrons were his passion. His studbook records 1210 crosses of which 462 new varieties were named and registered. At the heart of his hybridization programme were the rhododendron species. These were being collected by those remarkable plant hunters Forrest, Kingdon Ward, Wilson and Farrer on their expeditions to China and other countries in the Far East. Lionel de Rothschild left the legacy of a beautiful garden, enjoyed today by thousands of visitors, and a superb collection of rhododendrons. The present owner, Mr Edmund de Rothschild, continues with the hybridization programmes.

This was the background to the Gardens, at which I arrived in 1980 to paint 'a few azaleas'. It was to influence and entrance me then, and now.

My love for the species started very slowly. At first, it was natural to marvel at the prize-winning hybrid blooms, so spectacular with their intense colours and larger forms. But gradually, from the late 1980s onwards, the spring season was dedicated to searching for other species to paint. This extended into the summer as the species presented me with the surprising bonus of beautiful new foliage growth, with its many ranges of subtle colours and textures, and later, autumn brought seed capsules shaped as candelabras or woolly-coated. The species rhododendrons were all intriguingly different, full of character, and were to tempt my curiosity to the full.

To the newcomer, the natural progression on discovering the rhododendron species is towards the very large Subgenus *Hymenanthes* which holds the larger shrub or tree-like species in a single section, *Pontica*. They cover an immense range of form, colour and geographical distribution and, as we will see in the following pages, are the showiest species, much loved by the hybridists.

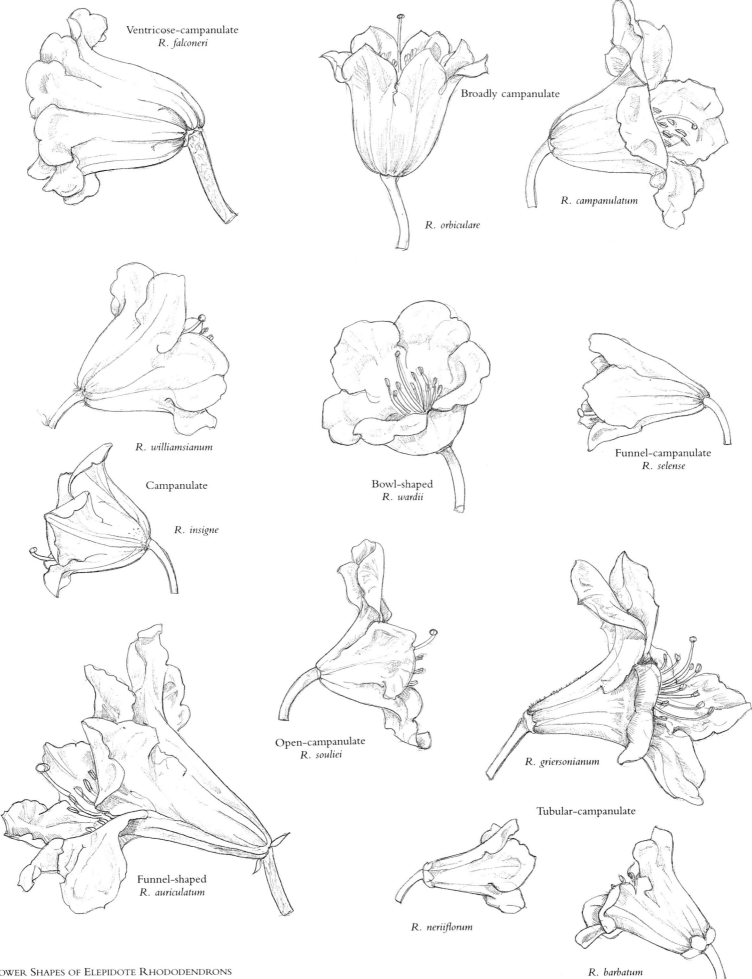

Ventricose-campanulate
R. falconeri

Broadly campanulate

R. orbiculare

R. campanulatum

R. williamsianum

Campanulate

R. insigne

Bowl-shaped
R. wardii

Funnel-campanulate
R. selense

Open-campanulate
R. souliei

R. griersonianum

Funnel-shaped
R. auriculatum

Tubular-campanulate

R. neriiflorum

R. barbatum

FLOWER SHAPES OF ELEPIDOTE RHODODENDRONS

Subsection **ARBOREA** Sleumer

(Series *Arboreum*, Subseries *Arboreum*)

Tony Schilling

*R*hododendron arboreum was first introduced to western cultivation in about 1810 by Dr. Francis Hamilton (later known as Francis Buchanan-Hamilton), with later collections being made by Nathaniel Wallich in 1821 and by Sir Joseph Hooker in 1849–51. Several trees from these early collections remain extant in British gardens to this day – one 7.6m (25ft) tall, white-flowered example from a Hooker introduction can be seen at Stonefield in Argyll, Scotland and other 140-year-old specimens at Lochinch, Scotland are also from Hooker seed.

Rhododendron arboreum (including its three subspecies and four varieties) is a very complex species and spans an amazing geographical range, from Kashmir in the west to South-West China (Yunnan and Guizhou) in the east as well as occurring in the Nilgiri Hills of southern India and the uplands of Sri Lanka. It is therefore hardly surprising that it is such a variable species, not only in leaf shape and indumentum, but also in flower colour. The indumentum colour varies from silver through fawn to deep cinnamon and ranges in texture from plastered to spongy, while the flowers, typically crimson, especially in the eastern half of its range, are frequently blood-red, rose-pink or white in the central section of its distribution pattern.

Named forms which have received awards from the Royal Horticultural Society include 'Goat Fell' (Award of Merit [A.M.] 1964), 'Rubaiyat' (A.M. 1968) and 'Tony Schilling' (F.C.C. 1974). There is also a wealth of hybrids containing *R. arboreum* 'blood' some of which were created over one hundred years ago (for example 'Lady Eleanor Cathcart') while others (such as 'Noya Chief' and 'Victoria Hallett') are relatively new to gardens. Various natural hybrids have also found an established place in horticulture, the most renowned being 'Sir Charles Lemon' (*R. arboreum* × *R. campanulatum*), which is traceable to one of Hooker's introductions.

I have a particularly strong personal affection for this majestic species of tree rhododendron, having spent the better part of three years of my life exploring within its forests, which range in altitude from 1675–3500m (5000–11,500ft). When I lived at Godavari in the Kathmandu Valley in central Nepal, over 25 years ago, I frequently admired its scarlet trusses in the dense forests of Phulchoke as I lunched on the terrace of my bungalow, which lay adjacent to the Royal Botanic Gardens.

At the foot of the hill the first flowers opened in late February, but as the season progressed one could follow the flowering sequence up the ridges, until by April the 15m (50ft) tall trees close to the 2750m (9000ft) summit were bedecked with scarlet, pink and white blossoms. Far away to the north one could admire the Himalayas stretching across the horizon, their cool-temperate flanks playing host to vast forests of *R. arboreum* and the many other plants which share its company.

Perhaps my most stunning encounter with this rhododendron was during the spring of 1983 when I trekked towards the Rupina La on the eastern flanks of the Gurkha Himal in central Nepal. There, the relatively untouched forests of *R. arboreum* were especially rich and beautiful in their variety, but their stature was almost beyond belief. I had during my numerous travels in the mountains of Asia seen many fine examples of rhododendron forest, but never before like this. The patrician trees reached heights of 24m (80ft) or more, and not just the occasional specimens, but in uniform ancient groves of mighty-stemmed magnificence. If ever there was an argument for a forest reserve providing protection to plant life, it is here in central Nepal where this noble species seemingly reaches its zenith.

The Nepalese call *R. arboreum* 'Lali Gurans' and it is recognized as their national flower. The Himalayan highlanders use the wood for turnery as well as for firewood and charcoal. The blooms, which are edible but sour of flavour, are used for the treatment of throat complaints, as a cure for dysentery, in poultices for the relief of headaches and as religious offerings in temples.

No other rhododendron species in the world has such a wide geographic distribution coupled with so many different vegetative characteristics. It has won the hearts and minds of many plantsmen over the years, including that doyen of twentieth-century planthunters, Frank Kingdon-Ward, who wrote: "…it would be *lèse majesté* to ignore this ancient member of a royal family".

Clockwise from left:

R. arboreum Sm.;

var. *album* Wall;

var. *roseum* (Lindl.) Tagg

shù xíng dùjuān "tree-like rhododendron"

a

a

a

g

b

gii

gii

ci

hiii

d

ei

gi

k

SUBSECTION **ARGYROPHYLLA** Sleumer

(Series *Arboreum*, Subseries *Argyrophyllum*)

SCOTT VERGARA

The Subsection *Argyrophylla* contains about 17 species and 6 subspecies, which are found in greatest concentrations in the Sichuan, Yunnan and Guizhou provinces of China. They are often found growing in woodlands at altitudes of 1200–3500m (4000–11,450ft). Generally, members of this subsection are shrubs or small trees with young shoots covered by a thin indumentum which falls off as they mature. Leaves are narrowly elliptic to oblanceolate with a thin, plastered, continuous indumentum of hairs on the lower surfaces, similar to the *Arborea* subsection. This indumentum is one of the most distinguishing features of the subsection. Flowers are usually five-lobed with a small calyx and are carried in loose trusses of 4–20. Colours vary from white, pink, rose and purple to lilac-purple, with scarlet and crimson notably lacking. Many are rare in cultivation, and some have never been successfully introduced.

Rhododendron adenopodum, *R. floribundum* and *R. hunnewellianum* (including its subspecies) were collected by Wilson during the early 1900s in Sichuan and surrounding areas. These species differ from the others in the subsection by possessing a more woolly indumentum. *R. adenopodum* is rather rare in the wild and has 6–10 flowers of pale rose, occasionally with purple spotting, in mid- and late spring. It is a hardy shrub, rounded and somewhat lax, growing to 3m (10ft).

Rhododendron argyrophyllum is probably best known by the A.M. clone 'Chinese Silver' of its subspecies, ssp. *nankingense*, which has good pink flowers and white indumentum on the new growth. Introduced by Wilson in 1904, although described in 1886, its name means "with silver leaves". This a widely grown and highly desirable plant. *R. argyrophyllum* ssp. *argyrophyllum* is the most common form as a shrub or small tree 1–12m (3½–39ft) in height, being somewhat variable in habit. It is hardy to −18°C (0°F) and blooms as soon as early spring. The long narrow leaves, smooth, dark green above with a thin, white, silvery to fawn indumentum beneath, are the unique characteristics of the plant. The flowers vary in colour from white or white-flushed-rose to pink, sometimes with darker pink spots or stripes, and are bell-shaped in lax trusses of 6–12 blooms.

Rhododendron insigne, whose name means "remarkable", truly is a remarkable plant and is often considered as the best of the subsection. Discovered by Wilson in 1903, this shrub grows slowly to 1.5–4m (5–12ft) with a rather upright habit. The very hard and stiff 13cm (5in) long, oblong-lanceolate leaves are a dark glossy green. The hard, thin indumentum, which varies from a tawny grey to a pale coppery sheen, is the most dramatic aspect of the plant. It is hardy to −18°C (0°F) and is somewhat rare in western Sichuan, where it grows in woodlands and on limestone bluffs. Flowering is in late spring and early summer with trusses of 8 to 16 bell-shaped, pinkish-white blooms marked with crimson flecks and external rosy-pink stripes. A little slow to start flowering, it is free flowering once it begins. The young bark is a salmon colour that turns grey with age. An outstanding characteristic of the species is long leaf retention, often over five years. The foliage characteristics carry through in many of the hybrids that have been made. The flower colour range is from white to pink and salmon-pink, to rosy-purple to a reddish-fuchsia.

The distinctive dark nectar pouches of *R. ririei* make it the most unusual flower of the subsection. It is intermediate between this subsection and Subsection *Arborea* in this characteristic. The broad, bell-shaped flowers in lax trusses of up to ten often open as early as late winter at the Rhododendron Species Foundation and Botanic Garden in Washington, U.S.A. in light purple to smoky-blue colours. It develops into a small tree, 3–13m (10–43ft) with an open upright habit. The bright matt green leaves often have a very thin to sometimes non-existent silvery-white to greyish indumentum on the lower surfaces. Wilson found it in the Mount Omei area of Sichuan in 1904 in forests and exposed areas. It is a very hardy species, tolerating temperatures down to −20°C (−4°F).

Superficially resembling *R. hunnewellianum*, *R. thayerianum* is a slow-growing shrub, 1.8–4.3m (6–13ft) with a compact, upright growth habit, which was collected only once, by Wilson in October 1910 in West Sichuan around 3050m (10,000ft) in woodlands. Bell-shaped, white-tinged-with-pink or pink flowers are produced in early and midsummer in compact trusses of 10–20 blooms. This late-flowering habit is a highly desirable characteristic for extending the overall bloom season and may prove useful in breeding programmes. The leaves, which persist for three to six years, are narrow, leathery and stiff with a very thin, plastered, buff-coloured indumentum. It is hardy down to −18°C (0°F), easy to grow and deserving of much wider use.

R. insigne (Hemsl. et Wils.)
bù fán dùjuān "remarkable rhododendron"

a

k

ci

b

d

ci

h*iii*

g*ii*

g

g*i*

l*i*

Subsection **AURICULATA** Sleumer

(Series *Auriculatum*)

Dr. Hugh Dingle

"A strange invisible perfume hits the sense." *Rhododendron auriculatum* in full flower in the summer garden is as seductive as Cleopatra, and many people will remember, as I do, searching widely for the source of the gorgeous perfume before finding it.

This is a noble plant, attaining perhaps 6m (20ft) in height and wider than tall. Its mid-green leaves, up to 30cm (1ft) long when growing well, are characterized by the auricled base and by the hairs, particularly on the midrib, found on the undersurface; the petioles are strikingly glandular-bristly. Its flowering season is usually late summer, the last of the genus to flower, and new growth comes at about the same time, the young shoots clad with startling red ribbon-like bracts. Before mid-autumn the conical flower buds are an impressive 7.5cm (3in), the outer scales like a clutch of rapiers, maybe with curved or crossed points. It is not a heavily furnished shrub, as it loses its old leaves soon after the new expand. It is said to roll its leaves tightly in frost.

In a large, loose truss, the frilled and fragrant flowers are white, often pure but sometimes creamy, with a green touch at the base of the tube. E. H. Wilson, who introduced *R. auriculatum* in 1900, described a rosy-red form, but P. D. Williams of Lanarth asked in the *Rhododendron Society Notes* of 1922 if anyone had seen the pink form: clearly he had not! He reported a form at Lanarth which had a crimson blotch in the tube.

It is a demanding plant! It must have space, wind-shelter, and carefully calculated exposure to sunshine: sun to ripen late growth and ensure flowering, but not so much as to scorch the flowers or the edges of tender young leaves. Peter Cox says that it is very vulnerable to drought. Siting this *prima donna* is controversial – some recommend dense shade, some full exposure. Ernest Wilson, writing of rhododendrons in Hubei Province, tells us: "I never met with . . . *R. auriculatum* . . . elsewhere than in woods and forests."

In the 1971 *Rhododendron & Camellia Yearbook*, *R. auriculatum* was said to be "particularly at home" above Windermere (Cumbria, Northern England), and also "very much at home" in Lymington (Hampshire, southern England), and those lucky enough to have seen in its prime the magnificent plant at Arduaine in Argyll will have no doubt that it can be at home in Scotland too.

Rhododendron chihsinianum is the only other species presently allocated to Subsection *Auriculata*. It was discovered on the Kwangfu Expedition of 1955, and shares the elongated, conical leaf and flower buds of *R. auriculatum*. There is a drawing of the plant in *Rhododendrons of China*, the American translation of Volume III of *Iconographia Cormophytorum Sinicorum*. Tam Pui-cheung in *A Survey of Genus Rhododendron in South China* gives the flower colour as red. It is not in cultivation in the U.K.

There are over 35 hybrids of which *R. auriculatum* is one parent: all large, late-flowering woodland plants. Although the leaves are usually auricled, the absence of hairs on the undersurface and the presence of glandular bristles on the petioles declares hybridity. *R. griersonianum*, *R. facetum* and *R. kyawi* bring cerise or carmine to the colouring of *R.* 'Aladdin', *R.* 'Dragonfly' and *R.* 'Leonore', while *R.* 'Polar Bear', *R.* 'Argosy', *R.* 'Isabella' and *R.* 'Lodauric' ('Iceberg') retain the incomparable asset of scent. The great plantsman Collingwood Ingram always recommended 'Isabella' with its exceptionally large, lily-like flowers, but the best known is undoubtedly *R.* 'Polar Bear' (*R. diaprepes* × *R. auriculatum*), which was bred in 1922 by J. B. and Roza Stevenson at Tower Court, not to create a masterpiece, but specifically to form a tall shelter belt on their north-east boundary!

R. auriculatum (Hemsl.)
er yè dùjān "ear-leaf rhododendron"

m*i*

g*ii*

g

a

b

b*ii*

d

h*iii*

e*i*

k *viii*

k*i*

l*i*

SUBSECTION **BARBATA** (Tagg) Sleumer

(Series *Barbatum*, Subseries *Barbatum*)

IAN W. J. SINCLAIR

*R*hododendron barbatum, with its tight, red trusses and plum-coloured bark, was the first "hardy" rhododendron species that I fell in love with, so much so, that I made my first angling priest from a section of one of its branches! Whether or not the trout appreciated this finer point is open to debate. My first sighting of this species in flower was in the wilds of Bhutan, in the spring of 1993. I had been fortunate to see *R. barbatum* at various different localities in Bhutan since my first visit in 1984; however, to finally find it in flower was wonderful. This fine, 3m (10ft) high shrub was growing in magnificent fir forest at 3140m (10,300ft) on the west side of the Tsele La, in Wangdi Phodrang District.

Subsection *Barbata* is extremely striking, whether or not it is in flower. When in flower, the tightly formed inflorescences open early in the season and vary from scarlet-crimson to pink. When not in flower, their most striking feature is the peeling bark, which is either plum or red in colour. Currently five or six species are thought to belong to this subsection, and there is much debate concerning the possible hybrid status of *R. erosum!* The other species in the subsection are *R. barbatum, R. argipeplum, R. exasperatum, R. succothii* and possibly *R. shepherdii.*

When *R. barbatum* was first introduced is not totally clear. By 1849 J. D. Hooker had collected it in Sikkim; however, Wallich had earlier found it in Nepal in about 1829, and it may have been introduced at that earlier time. Later seeds from an unknown source were also grown by Messrs. F. & J. Dickson of Chester, and the species is recorded as having flowered in their nursery from as early as 1848.

Rhododendron barbatum forms a shrub or small tree of 2–9m (7–29ft), the tree forms being generally upright. Both forms develop magnificent, peeling, plum-coloured bark. The buds are sticky and red. The upper surface of the leaves is more or less glabrous, the lower surface is almost glabrous but can have a few dendroid hairs and stipitate glands, mostly restricted to the midrib. The petiole varies greatly, from having stiff bristles of 6–10mm (⅕–⅓in) to being occasionally glabrous. The compact inflorescence has flowers of crimson-scarlet to blood-red with dark basal blotches, which act as ultra-violet attractants to pollinating insects. The flowers, in cultivation, bloom from late winter to mid-spring, and in the wild from mid- to late spring.

The distribution of *R. barbatum* ranges over North India (Uttar Pradesh, Sikkim, Bengal, West Arunachal Pradesh), China (South Xizang), Nepal and Bhutan. It is found in spruce, hemlock and fir forest, from 2100–3650m (6890–11,970ft).

Asked to recommend two hybrids out of the *R. barbatum* "stable", my choices would be *R.* 'John Holms' and *R.* 'Shilsonii'. Given an Award of Merit (A.M.) in 1957, *R.* 'John Holms' was bred by the Gibsons of Glenarn to commemorate their friend John Holms, who between 1908 and 1910 developed the gardens and Scottish Baronial Hall at Formakin at Langbank by Port Glasgow, as well as the Larachmhor garden at Arisaig during the 1920s and early 1930s. *R.* 'John Holms' is a hybrid out of *R. arboreum* × *R. barbatum*. The parent plant at Glenarn is now 9m (30ft) high, and is worthy of a visit, as is the magnificent garden itself, which has been restored and further developed by Michael and Sue Thornley since 1983. In 1962 a group of gardeners and botanists from the Royal Botanic Garden, Edinburgh, were invited to become involved in the overall management and practical running of the Larachmhor garden. To celebrate the memory of the gardens' founder and their own continued involvement, they planted a young propagule of *R.* 'John Holms' in 1991. This plant is still growing well at the time of writing, despite some minor browsing from a roe deer.

Rhododendron 'Shilsonii' was awarded an A.M. in 1900 and was bred by R. Gill, the Head Gardener of Tremough, who named it after the garden's owner. *R.* 'Shilsonii', a hybrid out of *R. thomsonii* × *R. barbatum*, is a very striking plant, forming either a very large shrub or a small upright tree. It has most of the good characteristics of both its parents – the bark is similar to that of *R. barbatum*, while the leaves are closer to those of *R. thomsonii*. The trusses are very fine, producing waxy, red flowers early in the season. The finest display that I have seen of this hybrid was at Muncaster in Cumbria where it has been planted in mass – early to mid-spring is when you should plan your visit.

R. barbatum (Wall. ex G. Don)

yìng cì dùjuān "stiff-thorned rhododendron"

a

k

ji

h

ci

hv

b

e

d

g

gii

SUBSECTION **CAMPANULATA** Sleumer

(Series *Campanulatum*)

MAJOR THOMAS LE M. SPRING-SMYTH

*R*hododendron *campanulatum* is of Himalayan origin and its arrival in Britain in 1825 makes it the second oldest introduction from India in the days of the old East India Company. *R. campanulatum* is found from Kashmir in the west through Nepal to Sikkim and Bhutan, but it is only since the 1950s, when Nepal ceased to be a closed country, that collecting seed there became possible for the first time.

Some of the original introductions were not notable for the colour and quality of the inflorescence and the leaf indumentum was often unexciting, so *R. campanulatum* was easily outshone by many of the exciting new species introduced in the early years of the twentieth century. This should no longer be the case.

A good form of this species grows as a well-rounded shrub or medium-sized tree. The open, campanulate flowers, some five to twelve or more to a truss, range in colour from lavender to pale mauve and pink to white. The leaves, ovate to broadly elliptic, are distinguished on the underside by a thick, rich golden-brown indumentum, akin to a layer of velvet.

This rhododendron, which prefers an open habitat, is at its best growing on rising ground so that at all times of year the light of the morning and evening sun can fall on the leaves, giving the plant a stunning golden glow when seen from below.

There is a clone, *R. campanulatum* 'Knaphill' with lavender blue flowers and pale fawn indumentum, exhibited by Mr. Lionel de Rothschild, which received an Award of Merit (A.M.) in 1925.

There are two subspecies. One of these, *R. aeruginosum,* grows in Sikkim and Bhutan and is distinguished for its young, glaucous, blue-green foliage. The other species, *R. wallichii,* grows in Nepal and through Sikkim and Bhutan to Assam. In most respects it is very similar to *R. campanulatum* except for the indumentum, which is extremely sparse and of no distinction. It is a shrub for the botanic garden rather than anywhere else.

In addition to 'Knaphill', two other *R. campanulatum* clones have gained A.M.s: 'Roland Cooper' and 'Waxen Bell', both raised at the Royal Botanic Garden, Edinburgh.

Perhaps because some introductions were less than inspiring, *R. campanulatum* has not generally been the hybridizer's first choice. However, when crossed with that splendid species *R. fortunei*, we have 'Susan', which carries deep lavender to almost blue flowers and grows large in time but without a well-coloured indumentum. It earned an A.M. in 1930, a First Class Certificate in 1954 and an Award of Garden Merit in 1984. Another possibly natural hybrid, 'Sir Charles Lemon', has already been mentioned in Subsection *Arborea*.

In the course of seed collecting in East Nepal over seven months, from the autumn of 1961 to the spring of 1962, I walked up the valley of the Mewa Khola until at dusk I left the forest and came upon a wonderful southward-sloping alpine meadow with grazing yaks just below the village of Topke Gola at 3860m (12,700ft). The next morning, 25th October, my birthday, revealed compact bushes of *R. campanulatum* (and *R. campylocarpum*), all about shoulder high and covered in seed capsules. Through the whole site flowed narrow rivulets of melting snow water from the peaks above. The combination of grass, rocks, water, the remains of primula, gentian, meconopsis, countless other plants and the rhododendrons was an unforgettable sight that was all the more heightened by the aromatic smell of dwarf rhododendron leaves crushed underfoot.

A return to Topke Gola on 24th January 1962 found the area under 90cm (3ft) of snow, the village deserted and a night-time temperature of −10°C (14°F). The leaves of *R. campanulatum* were tightly rolled like cigarettes, but in my shirt sleeves in the morning sunlight I collected more seed. I shall always remember *R. campanulatum* in the perfect setting of that wild, mountainside garden, tended by the grazing yaks.

R. campanulatum blue & white forms (D. Don)
zhōng huā dùjuān "bell-flowered rhododendron"

SUBSECTION **CAMPYLOCARPA** Sleumer

(Series *Thomsonii*, Subseries *Campylocarpum* and Subseries *Souliei*)

DAVID KNOTT

*C*ampylocarpa is only a relatively small subsection. Its four species, however, are among the most attractive and horticulturally important species in the genus.

Rhododendron souliei is perhaps one of my favourite species, with exquisite, saucer-shaped flowers that are pink in bud, opening white-flushed-pink fading to white, while the attractive glabrous foliage provides the perfect foil to this display. It has a reputation for being a difficult plant to grow, and seems to prefer the drier, colder gardens. At Dawyck Botanic Garden in the Scottish Borders, where minimum temperatures of -20°C (-4°F) are not uncommon and it is relatively dry, it appears to thrive and rewards us every May with its superb flowering.

I was therefore not surprised, on a trip to western Sichuan in 1991, to see *R. souliei* growing in full exposure at 3500m (11,500ft) on the Zheduo Shan, north-west of Kangding. It was first collected and described in 1895 by the French missionary and plant collector Jean André Soulié when he was based in Tatsienlu (Kangding), in what were then the Tibetan border regions. It was introduced to cultivation by Ernest Wilson in 1903 when he was collecting for Messrs J. Veitch and again on his two subsequent expeditions for the Arnold Arboretum.

Rhododendron souliei has been awarded three First Class Certificates (F.C.C.) from the Royal Horticultural Society – the first was awarded in 1909 to a form with pale rose flowers, shown by Messrs J. Veitch; the second in 1936 to the clone 'Exbury Pink' with deeper pink flowers, from Exbury; and the third to the clone 'Windsor Park' with white-flushed-pink flowers, from the Crown Estate Commissioners, Windsor. It has been used by the hybridizers, but I am afraid none of their efforts, in my opinion, improve on the species!

Rhododendron callimorphum was found by George Forrest on the Shweli/Salween divide, in western Yunnan, and introduced by him in 1912. In cultivation, *R. callimorphum* var. *callimorphum* forms neat, rounded bushes, 2–3m (7–10ft) high and the same wide, producing during mid- and late spring flowers that are an attractive pale or deep rose, sometimes with a deep crimson blotch at the base. Like *R. souliei*, it also prefers drier conditions at the rootball but it is not nearly so hardy. *R. callimorphum* var. *myiagrum* is a white variant, discovered by Forrest.

Rhododendron campylocarpum ssp. *campylocarpum* was discovered by Sir Joseph Hooker in East Nepal during 1848 and, it is said, of all the rhododendrons that he saw on his famous expedition, *R. campylocarpum* was the one he most admired. It is easy to see why with its attractive pale to sulphur-yellow flowers and compact habit, producing a bush 1–3m (3–10ft) tall and round. There are, however, two different forms in cultivation – Hooker's original compact introduction and another, laxer-growing form, Hooker's form, which received an F.C.C. in 1892. Both forms prefer mild, moist gardens and have been used extensively by the hybridizers – one of the nicer old hybrids is 'Logan Damaris', now unfortunately quite susceptible to rhododendron powdery mildew.

Rhododendron campylocarpum ssp. *caloxanthum* was discovered by Reginald Farrer and Euan Cox in Upper Burma during 1919, and plants raised from this collection received an Award of Merit (A.M.) when shown by Exbury in 1934. In cultivation, *R. caloxanthum* has proved to be a most attractive plant, with cream to yellow flowers that are often orange in bud and forming a bush 1–2m (3–6½ft) high and as much wide. Like *R. souliei*, it requires drier conditions at the rootball to grow well.

Rhododendron wardii was first described and introduced from specimens collected by Frank Kingdon-Ward – after whom it is named – and by George Forrest in 1913, in North-West Yunnan. It does, however, occur over a widespread area of North-West Yunnan, South-West Sichuan and South-East Xizang, and there are many different forms in cultivation. The best of these forms, which have gained seven Awards of Merit, and their offspring are among the finest yellow rhododendrons for general cultivation. In cultivation, it grows into a bush 1–10m (3–32ft) tall and requires an open situation to remain compact. The flowers are saucer-shaped, in various shades of yellow, some with and some without a crimson blotch.

R. souliei Franch. var. 'Exbury Pink'
bái wǎn dùjuān "white-bowl rhododendron"

a

b

d

ei

ji

hv

g

hv

ci

oi

gii

h

k

SUBSECTION **FALCONERA** Sleumer

(Series *Falconeri*)

MICHAEL & SUE THORNLEY

The Subsection *Falconera* is treated in the 1980 Edinburgh Revision as follows: 1. *R. rothschildii*, 2. *R. semnoides*, 3. *R. basilicum*, 4. *R. rex* (a. ssp. *rex*, b. ssp. *fictolactum*, c. ssp. *arizelum*), 5. *R. preptum*, 6. *R. galactinum*, 7. *R. coriaceum*, 8. *R. hodgsonii*, 9. *R. falconeri* (a. ssp. *falconeri*, b. ssp. *eximium*), 10. *R. sinofalconeri*. A few of these species are grown quite widely in British gardens, but numbers 1, 2, 5, 6, 7 and 10 are not easily found except in very good and relatively extensive collections.

The most conspicuous feature of most of these plants is their sheer nobility – the whole scale of the plant, foliage and inflorescence. It is not surprising that rhododendron enthusiasts have almost universally adopted the sobriquet "The Big Leaves" to indicate the two subsections *Grandia* and *Falconera*. In comparison, all the rest are dwarfs.

Take tall *R. eximium* (3–9m/10–30ft) for instance, which is clothed from head to foot in creamy trusses that almost touch the ground. With the sunlight catching the cinnamon-coloured indumentum on its leaves, the sight of this excellent rhododendron takes the breath away, both by its aloofness and beauty. Then, in summer, a much more intimate *R. eximium* reveals its own new growth of soft rabbit's ears.

In contrast, *R. hodgsonii* is raffish and gregarious, an extrovert with wide-open habit, displaying an extraordinary, shaggy, mauve trunk and well-manicured, silvered foliage. But more than any other feature it is the flower that catches the eye, like an outrageous buttonhole. Early each spring the small, electric-pink trusses, with golden dotted stigmas, startle the eye although their colour all too quickly fades to washed-out-rose. This quality – for it can hardly be called a fault in a tree which is so useful in its native lands of Bhutan, Sikkim and Nepal – was perpetuated when *R. hodgsonii* was crossed with *R. sinogrande*. The result was the even more eccentric *R.* 'Ronald', with strawberry-ice, football-sized trusses and large black, languorous, metallic leaves. It was given an Award of Merit in 1958.

Rhododendron basilicum and its near neighbour *R. rothschildii* are altogether more stolid plants with their feet firmly set in the ground, clad in flat-petioled leaves which are arrayed like armoured plates. In keeping with their habit the flowers are more reticent, to the point of almost being shy in the case of *R. rothschildii*. The odd man out in the subsection is *R. fictolacteum* with its variable foliage that ranges from the almost retarded at Brodick Castle on the Isle of Arran, Scotland, to leaves of reptilian splendour on the Isle of Colonsay, Scotland. Even more than the other members of the subsection, *R. fictolacteum* requires good protection and a favourable site in which it can slowly develop and eventually produce its delicate flowers.

Towering above all the rest, however, is *Rhododendron falconeri*. It came here to the garden at Glenarn in Scotland as seed first collected by Joseph Hooker on his 1849 expedition to Sikkim. This old tree sits like a huge buddha, a reincarnation from another continent and time, whose vast flesh-coloured limbs have grown into an oriental pattern against the grey Scottish sky. Later its own seed was broadcast in the Upper Glen where a group of *R. falconeri* now stands, the rain rattling on the wide canopy of their leaves. Hold a single leaf in your hand – on its upper surface it is the colour of dark green leather and the jungle. Turn the leaf over to reveal another world: a miniature tree with its branch-like structure of yellow veins standing on a background of rusty indumentum.

Every two or three years the venerable tree covers itself in waxy ivory flowers, with a purple blotch lying deep inside each corolla. The abundance and length of time the flowers stay in their prime, which can be up to a month, seems almost life-threatening, especially in the dullness of late summer when the seed capsules split open in contorted, ragged heads that check the growth of the new foliage. But the cycle slowly continues.

It does not seem surprising that *R. falconeri* and *R. hodgsonii* hybridize freely in the wild. At Glenarn the Gibson brothers crossed *R. falconeri* with the yellowest of *R. macabeanum* to create a marriage of near perfection between the king and queen of the rhododendron world.

R. falconeri Hooker (after Falconer, 1808–1865, Supt., Saharanpur Gardens, India, in 1832)

kiii

a

b

hiii

g   *ei*   *d*   *gii*   *ci*   *li*   *n*   *p*   *n*

Subsection **FORTUNEA** Sleumer

(Series *Fortunei*)

John D. Bond

Without doubt my favourite rhododendron hybrids are to be found amongst the best of the × *loderi* group. A strange beginning to a note on rhododendron species in general and the *Fortunea* subsection in particular? Not so, for at least half of the credit for the Loderis has to be given to *R. fortunei*, the species which gives its name to this subsection, and half to *R. griffithianum*, also placed in this subsection. Not only are these superb, comparatively easy and reliable large shrubs, they also make excellent parents. Flowering time for members of the *Fortunea* subsection extends from mid-winter until late summer.

My problem is where to begin this note and I have chosen the beginning of the year when, if the weather is reasonable, *R. oreodoxa* will add great cheer to the winter scene. A neat-foliaged, fairly small-flowered species which eventually attains 4.7–5.5m (15–18ft), this pretty shrub will flower freely, sometimes almost too freely, having reached a reasonable age. It is, unfortunately, frequently frosted in flower. I advocate deadheading all rhododendrons as soon as the flowers have faded, particularly with species which tend to overflower, and *R. oreodoxa* is one of several species in this subsection which overdoes things with age. A choicer species, the closely allied *R. fargesii*, flowers a month or so later – the deeper pink and larger flowers are produced amongst excellent foliage. This species is of less stature than *R. oreodoxa*.

Rhododendron sutchuenense and the closely allied *R. praevernum* also flower early, late winter to early spring in fact, and again we take a great risk with frosts. Beware when searching for plants of these species, for there are many inferior forms and seedlings to be seen in gardens and no doubt offered in nurseries! The best forms, however, are superb and there are few finer sights than a well-flowered, mature specimen of *R. sutchuenense*, which may well have attained 6–7.5m (20–25ft). Flower colour varies from white to a good deep pink. Even more exciting is the allied *R. calophytum* with large trusses of white or clear pink flowers set among superb lance-form foliage, which, in a healthy young plant, can extend to 38cm (15in) in length. The overall outline of this species is dome-like and will eventually attain 4.7–5.5 × 5.5m (15–18ft × 18ft).

There is yet another beautiful, early-flowering species: *R. vernicosum*. At least in its best forms it is beautiful, for there is great variation in leaf and flower quality and this species is infrequently offered in the trade.

And so we find ourselves in mid- or even late spring, when we may well be admiring a tall, perhaps 7.5–9m (25–30ft) specimen of *R. fortunei*, covered with mauve-pink flowers of great quality and beauty. Again a warning, there are many inferior plants of this species to be seen in gardens, most of which are seed-raised, often self-sown plants which are just not true to name. Open-pollinated species in gardens rarely produce typical plants from seed and some species are more prolific than others. *R. fortunei* is one of them! However, the species, when true to name, is superb and well worth hunting for. The name commemorates the famous plant hunter Robert Fortune, to whom we owe much. A true plant of the similar *R. houlstonii* is also worth seeking, but it is *R. discolor* which is so very useful in extending the season into early and midsummer. *R. discolor* will attain 6–9m (20–30ft) fairly rapidly, and there is no finer sight when it is covered with its white or pink flowers. These three species are clearly closely allied and should find a place in all large gardens. Their habit is leggy, almost tree-like with age, which is an asset, for a grove of bare rhododendron stems has great architectural value in the garden. These three species provide an added bonus: outstanding delicate scents.

Lastly we must consider the species depicted so beautifully here, the lovely *R. decorum*, which is again scented and again varies from clean white to pale pink. While there is less variation in this species' flower, foliage or habit, there is considerable variation in hardiness – so again beware! Having reached a reasonable age, ten years or so, *R. decorum* will flower freely in late spring and early summer. In fact, it tends to be another species to overdo things, so deadhead as soon as the flowers have faded. The pure white *R. diaprepes* is best described as a larger-leaved and -flowered edition, which extends the season into mid- and late summer.

There are other species in this subsection; however, those described above are the best and are certainly important members of this great genus, *Rhododendron*, which has almost always been to the fore in my journey through horticulture, not only in the establishments where I have worked, but also in the innumerable gardens that I have visited over the years.

R. decorum Franch.
dà bái dùjuān "great white rhododendron"

a

ci

hiii

d

ei

li

g

giii

gi

j

k

SUBSECTION **FULGENSIA** Chamberlain

(Series *Campanulatum*, Part) (Hooker)

ALAN CLARK

I first became aware of this wonderful species in the spring of 1979 at Lamellen in Cornwall. Walter Magor was showing me around his garden, and from one of the higher points my eyes were drawn to a plant with blue leaves. Even from a distance of 300m (1000ft) the metallic sheen stood distinct from the various shades of green sported by other rhododendrons. At this point in time I had only recently become infected by the "rhody bug" and my brain raced, "Is it *R. thomsonii, R. clementinae* or *R. campanulatum aeruginosum?*" Even with my limited knowledge it did not quite fit any of these species. Eventually we stood close to the plant. I was invited to turn a leaf over, which revealed the most exquisite, silvery, moss-like indumentum and purplish-grey bark on all but the thinnest branches. Now totally confused, I enquired, "What colour are the flowers?" – "Crimson," came the reply. My misery soon ended, "You are looking at *R. fulgens,*" my host explained.

This species was introduced by J. D. Hooker from Sikkim in 1849–50, flowering for the first time at Kew in 1862. It has since been collected often in Nepal, Bhutan, North-East India and Xizang (Tibet). I was fortunate to find a single plant growing at 3900m (12,800ft) above the village of Mera in eastern Bhutan in October 1990. On this memorable day we trekked from the village heading north-east, at first through open pasture – *Juniperus recurva* dominated the landscape, carved into strange shapes from the attentions of hungry yaks. From above, the valley appeared set for a game of chess of epic proportions; the bright light and complete silence intensified the feeling that only the players were needed to complete the illusion. Suddenly we entered the *R. thomsonii* zone, and for miles across the slopes grew the species by the thousands, in spring a lava flow. This was followed by open moorland with occasional dwarf species and then, on the final approach to Meratot, on the lee side of a ridge in the company of *R. flinckii, R. bhutanense* and *R. wallichii,* stood the lone plant. An extensive search revealed no others.

When first described, *R. fulgens* was placed in the *Campanulata* subsection, although its indumentum type and flower colour clearly place it apart. Currently there are only two other species within Subsection *Fulgensia. R. miniatum,* collected by Ludlow and Sherriff in South-East Xizang is not believed to be in cultivation and may be a form of the third member, *R. sherriffii.* This species, as the name suggests, was also collected by Ludlow and Sherriff, near the Chayal river in Xizang. From a horticultural view, its affinity with *R. fulgens* is doubtful, the only common feature being the flower colour. *R. sherriffii* has the most beautiful, loose truss of crimson flowers. The leaves are dark green above and in some forms the purple colour of the petioles extends into the leaf. The undersurface is covered in a dense, suede-like indumentum. Two superb plants of the species can be seen at Wakehurst Place, Ardingly, Sussex, in England; these are now 2m (6½ft) tall.

R. fulgens Hooker f. "shining"

a

li

n

gii

kiii

hiii   eiii   b   oi   d   ei   ki

Subsection **FULVA** Sleumer

(Series *Fulvum)*

Dr. Robert H. L. Jack

Within *Rhododendron* subgenus *Hymenanthes* Section *Pontica, Fulva* is a small distinct subsection containing only two species: *Rhododendron fulvum* (Balfour f. & W. W. Smith 1917) and *R. uvarifolium* (Diels 1912). Both were discovered and introduced by the plant collector George Forrest – *R. uvarifolium* in 1904, *R. fulvum* in 1912 from western Yunnan, province of China. Although his was the first introduction, both species were subsequently found by other collectors in western China, South-East Xizang (Tibet) and North-East Burma.

Rhododendron fulvum (the name means "tawny") is a delightful plant in form, flower and foliage although not widely grown in gardens. It can attain the stature of a small tree, up to almost 6m (20ft), or at least of a large shrub. For this reason it is usually grown in woodland gardens. It is hardy and flowers freely once established, but as a young plant it can suffer from bark-split after spring frost, which further underlines its suitability as a woodland plant. In this environment, the risk of damage from sudden spring frosts is reduced by the protection afforded by the top cover of woodland trees.

The flowers in mid-spring are pink but can range through pale pink to white, and they usually have an attractive dark crimson blotch at the base of the corolla. The flower truss is well-shaped, with 10 to 18 or even 20 flowers.

It is, however, the foliage that gives *R. fulvum* its unique quality. Each leaf is relatively broad and dark glossy-green above, with fawn- or cinnamon-coloured indumentum underneath. Moreover, this has the texture of suede leather, both in appearance and to the touch. When new leaves emerge in early summer they have at first a silvery-white covering. This gradually acquires its fawn or cinnamon colour as the leaves expand. The upper surface quickly becomes smooth while the green colour darkens. On the lower surface the indumentum darkens and remains. The leaves last two to three years, with the indumentum becoming a darker shade as each year passes. They are quite sturdy but lift and turn easily in a breeze, sufficiently to show the colour of their undersides. This is attractive in any season, but is particularly so when the flowers are open. The combined effect of a plentiful display of pink flowers, for it is free-flowering, set off by the dark green, glossy foliage is delightful. This in turn is emphasized by the contrasting, strong cinnamon undersurfaces, which show as the leaves move in a breeze. Such a combined display is special to *R. fulvum* alone.

In winter the leaves – although tough and not easily damaged by frost – nevertheless roll into tight, down-hanging quills whenever cold weather starts. There they stay, giving the plant a desperately miserable appearance, until the temperature rises whereupon they unroll and lift again.

Rhododendron fulvum is altogether a species of the greatest charm, and should be planted much more widely. When *R. fulvum* was exhibited to the Royal Horticultural Society by the Hon. H. D. McLaren, Bodnant, in 1933 it received an Award of Merit. Curiously it has not been used for hybridizing. Only one example seems to have been recorded, using *R. lacteum* pollen – the result, *R.* 'Hilde', does not appear to have been distributed and is otherwise unknown.

Likewise, *R. uvarifolium*, with its pink or white flower and silvery indumentum, although of smaller stature (usually of bush form), has not attracted hybridists' attention either. It is, however, a worthwhile garden plant in its own right. There is a variety of *R. uvarifolium* called *griseum*, meaning "grey", which has an especially ash-white indumentum that is smooth and silky in texture. Unfortunately it is rarely seen in cultivation.

There are several good specimens of *R. fulvum* growing in the Royal Botanic Garden, Edinburgh, one in particular, which happens to be from Forrest's original collection in 1912, F.8989, is well situated at the lower edge of a woodland area.

One spring afternoon, seeing that it was already in flower, I hastened nearer for a closer look. Almost there, rounding a thick bush, I nearly collided with a young couple holding hands and gazing into each other's eyes. Unable to retreat I found their gaze turned on me. To redeem the situation I said boldly, "Isn't this tree a wonderful sight? Pink flowers against a blue spring sky. And have you noticed the lovely cinnamon colour under the leaves?" In truth they had not seen anything. "All the way from western China," I continued, now starting to retreat. "Absolutely delightful," I added getting further away. This last fell on deaf ears, for they were once again totally lost in each other's gaze. A love-lorn pair indeed, and under my target *R. fulvum* of all places! Still, what a wonderful setting and from that starting point I am sure they got married and are living happily ever after.

R. fulvum (Balf. f. et W. W. Sm.)

liàn guǒ dùjuān "sickle-capsule rhododendron"

k

j

a

b

gii

g

ci

hiii

d

ei

o

gi

SUBSECTION **GRIERSONIANA** (Davidian ex) Chamberlain

(Series *Griersonianum*)

MARIANNA KNELLER FOR DOUGLAS BETTERIDGE

Douglas Betteridge has worked at Exbury Gardens for 40 years. His assistance in bringing me the species from the garden has been invaluable. Somehow, he managed to find time to accommodate my list of requirements for the season, in between showing guests around the garden, attending rhododendron shows, helping clear another bed for planting and a hundred and one other things that head gardeners have to do in a day. This article and the accompanying illustration are my thank you to him, as I know *R. griersonianum* is one of his favourite species.

It was a hot day in midsummer when Dougie came into my studio, carrying a spray of the most glorious glowing red rhododendron. At that late stage of the flowering season, species are rather scarce, so it was natural for me to assume that these flowers, because of their colour and size, were hybrids. However, Dougie threw another Latin name at me and said that in his opinion it was the "best of the lot". He went on to tell me that many award-winning hybrids owed their existence to the gene pool created from this species – *R. griersonianum*.

The day was late, and it was time to travel home, so I placed the blooms in my cool box, noting as I did so that they had already passed their best; indeed, next morning the flowers had dropped, leaving only the ovaries and the long red pistils.

It was to be a whole year before the notes on my calendar indicated it was time to alert Dougie that *R. griersonianum* might be flowering. Armed with his directions I set off towards the Main Drive, then turned into Lovers' Lane, which would take me to Witchers' Wood, called thus after the family of New Forest gypsies who camped there in the last century. The third sward on the right would lead me to a path where I would see an old pine tree to the east. "You can't miss it," Doug had said, "and north of that, at about a hundred feet, is the griersonianum."

Any doubt I had that my memory had exaggerated its colour and form were banished, for I was looking at the original planting of some 60 years or so ago – the parent and grandparent plant of some of the greatest prize-winning hybrids of this century. It was truly magnificent, a huge circular bush about 4m (12ft) high, with wayward strands of golden honeysuckle entwined around some of the older straggling branches. The profusion of flowers of such a powerful, vibrant, geranium-scarlet red colour spread from the lower branches among the grass and wild flowers to the upper reaches above. The plant was entirely and absolutely spectacular.

Using my secateurs I chose a lovely flower and placed it in a large polythene bag. This I blew up like a balloon, secured it with an elastic band and safely transported the plant back to the studio, content to have another beautiful species to paint.

Rhododendron griersonianum shares with *R. auriculatum* very distinctive flower and foliage buds with very long, tapering bracts, extending well over the buds, as well as a late flowering season. No other species has these characteristics. There, however, the similarity ends and *R. griersonianum* has the status of being the only species in its subsection.

It originates from the open pine forests of western Yunnan, where it was discovered by that remarkable plantsman and explorer George Forrest in June 1917. He gave it the number F. 15815, and noted its exceptional vibrant colour, and, remarkable for a red rhododendron, its fragrance. It was to be the most important species collected by him; within a few years of its discovery hybridizers were enjoying its unique contribution to the development of spectacular new hybrids, for plants were in flower in England by 1923. The next year it was awarded the First Class Certificate (F.C.C.).

This widespread enthusiasm for using the exciting new species with its incomparable colour had the result that when the R.H.S. "Stud Book" of Rhododendron Hybrids was published in 1969, it recorded an unequalled roll-call of 154 crosses with *R. griersonianum* as a parent. Since then, over 50 have had an Award of Merit (A.M.) or F.C.C. awarded by the R.H.S. Breeding achievements since 1969 are difficult to assess, but of course they have not come to an end; some of the hybrids produced have themselves been used as parents. Here at Exbury there have been many notable successes, including the award plants 'Damozel' (A.M. 1948), 'Daydream' (A.M. 1940) and 'Diva' (A.M. 1937). A complete display of award plants from all hybridists would exhibit colours ranging from very dark red through orange-reds into salmon and lighter pinks. The flowering season of the entire group would be well over three months, with a very strong representation towards the end of the rhododendron season.

R. griersonianum Balf. et Forr. (after R. C. Grierson, Chinese Maritime Customs at Tengyueh, friend of George Forrest) zhū hóng dà dùjuān "vermilion large rhododendron"

a

k

h*iii*

b

e*i*

c*i*

j

j*i*

g*ii*

g*ii*

g*i* and l*i*

SUBSECTION **GLISCHRA** (Tagg) Chamberlain

(Series *Barbatum*, Subseries *Glischrum*)

NIGEL PRICE

*T*he most distinguishing characteristic of this subsection is the presence of glandular hairs, especially on young growths, which can make deadheading a particularly sticky operation; *glischra* actually means "sticky". Eight species and subspecies make up the subsection, and while some are really only suitable for the connoisseur's collection, many are very attractive and well worth growing for foliage and flower. They are all reasonably hardy and, with the right conditions, should succeed in most sheltered woodland gardens.

Rhododendrons of this subsection come from some of the most prolific rhododendron areas: Yunnan, Upper Burma, South-East Xizang (Tibet) and Sichuan, and all but two were first introduced by George Forrest between 1914 and 1929. Frank Kingdon-Ward collected the delightful *R. recurvoides* and the rather less exciting *R. vesiculiferum* in 1926.

Rhododendron glischrum varies much in height and habit but is most often seen in gardens as a large, dome-shaped shrub, some 2–3m (6½–10ft) tall. The inflorescence, typical of the subsection, is a rather loose truss of bell-shaped flowers, ranging in colour from a very pale pink to purple. Many forms have a dark basal blotch, others are freckled with red spots, the best rather like a large and exotic Italian ice-cream. *R. glischrum* ssp. *rude* is very similar, the only real difference being the upper leaf surface, which is hairy.

Rhododendron glischrum ssp. *glischroides* is a far better plant altogether: its flowers are larger, and the leaves are infinitely more attractive, having deeply impressed veins above and petioles clothed with bright red hairs, and its habit is more compact. Flowers are produced in early and mid-spring, usually two or three weeks earlier than *R. glischrum*, and in the wild they vary in colour from white to deep rosy-pink, most having a prominent, dark red basal blotch. Forrest discovered *R. glischrum* ssp. *glischroides* in Upper Burma and Yunnan at a height range of 2740–3355m (9000–11,000ft), where it clings precariously to steep, boulder-strewn hillsides and cliff ledges.

The best foliage plant in this subsection is undoubtedly *R.*

crinigerum, which has superb, dark green, glossy leaves with deeply impressed veins and thick, cinnamon-coloured indumentum on the underside. The best forms seen in our gardens today were discovered and introduced by other collectors; indeed, George Forrest was less than complimentary when describing his own introduction of 1914. Flower colour varies from white to dark pink, the most attractive form having darker bands running from the base along the mid-lines of each lobe.

Another species with very fine foliage is *R. recurvoides*, although there is some doubt as to whether this should be in the *Glischra* subsection at all. In appearance it is very similar to *R. roxieanum* of the *Taliensia* subsection, and this is where it should remain according to some authorities. Collected in 1926 from steep scree slopes in North-East Upper Burma by Kingdon-Ward, he describes it as a dwarf, not more than 30cm (12in) high with short, thick stems ending in rosettes of narrow leaves, "completely overwhelmed with blossom". In cultivation, *R. recurvoides* reaches a quite respectable 90cm–1.2m (3–4ft), with narrowly lanceolate leaves, recurved margins, and a dense layer of cinnamon-coloured indumentum covering the lower surface. The flowers are white or pink with dark red spots.

Lastly, if I had space for just one plant from this subsection it would be *R. habrotrichum*. Distinguished by its dark red, bristly stems, hairy leaf margins and almost funnel-shaped flowers, this species forms a neat, medium-sized rounded shrub up to 3m (10ft) high. Flowers appear during mid-spring and vary from white to deep pink with a purple blotch, the darker colours looking particularly luxuriant against the deep green leaves and red branchlets. It was originally collected by Forrest and later by Reginald Farrer and Kingdon-Ward.

R. glischrum ssp. *glischroides* Tagg. et Forr.
niān máo dùjuān "sticky-haired rhododendron"

k

a

ci

lii

hv

b

d

gii

o

e

SUBSECTION **GRANDIA** Sleumer

(Series *Grande*)

AMBROSE CONGREVE

The *Grandia* subsection includes some of the most remarkable plants in the entire genus. They are distinguished by their size – both the plants and their leaves assume truly "grand" proportions. Often likened to elephants, they are magnificent, particularly when seen from afar.

The superb species *Rhododendron sinogrande* was discovered in 1912 by George Forrest on the Shweli-Salween Divide at 3355m (11,000ft), on its western flank. Seeds were sent back to Britain under Forrest 9021.

Rhododendron sinogrande occurs also along the mountains between the Salween and the eastern section of the Irrawaddy, to further north where it has been seen east of the Salween, at about the same latitude. Later, Forrest sent seeds from the area where Burma, China and Tibet meet. Ideally, seek the progeny of these plants as they must be hardier than the original introduction.

The young shoots arise in plumes of silver-grey from carmine tubes and make an arresting sight in the woodland. They should be well sheltered from the wind and the early morning sun and should have plenty of room. These growths develop into dark green glabrous leaves, silvery-grey beneath, which can be up to 60cm (2ft) long by 30cm (1ft) or more wide, thus displaying the largest foliage of any rhododendron except perhaps the tender *R. protistum (R. giganteum)*.

The flowers are produced in mid-spring, in large trusses 30cm (12in) high, with 20–30 blooms 6–7.5cm (2½–3in) long, ranging from cream to yellow, often with a crimson blotch at the base.

Rhododendron sinogrande was exhibited by G. H. Johnstone on 9th March, 1926, when it received a First Class Certificate (F.C.C.), having already received an Award of Merit (A.M.) in 1922. Mr. Lionel de Rothschild, probably the cleverest hybridist this century, who had created in some ten years the most spectacular garden in Europe at Exbury, must have at that time obtained one of the flowers of *R. sinogrande* and pollinated a plant of *R. falconeri*. The latter had been introduced by Colonel Sykes in 1830, and had received an A.M. in 1922 when exhibited by Messrs. Gill of Falmouth. The progeny, which Mr. de Rothschild named 'Fortune' – in my opinion the finest hybrid ever produced to date – then received a F.C.C. as early as 1938. This was a remarkably quick turnaround for *R. sinogrande*, which had only been raised from seed in Britain in 1912.

Mr. Gibson of Glenarn in Scotland crossed *R. hodgsonii* of Subsection *Falconera* with *R. sinogrande* to produce 'Ronald', which received an A.M. in 1958. One of the Glenarn plants is now a magnificent tree. In 1936 Mr. E. J. P. Magor of Lamellen, father of Major E. W. M. Magor, exhibited *R. 'Lacs'*, the result of using *R. sinogrande* as the pollinator of *R. lacteum*.

When asked to choose a species about which to write, I selected *R. sinogrande* on account of its outstanding qualities in the *Grandia* subsection, and also because the first and largest of many plants of *R. sinogrande* growing at Mount Congreve today was a gift from Lionel de Rothschild around 1930.

Other species included in Subsection *Grandia* are *R. grande, R. macabeanum, R. magnificum, R. montroseanum, R. protistum* (including var. *giganteum*). *R. pudorosum, R. sidereum* and perhaps the imperfectly understood *R. wattii. R. grande* has fine flowers, ranging from cream and pale yellow to pink and deep rose with purple blotches and spots, and is occasionally scented. Unfortunately, this species flowers early, from early spring, and is sensitive to frost. *R. macabeanum* is a particularly good and fairly hardy species. Buds start as attractive red candlesticks with handsome silvery-grey growth, but it is the contrast of its large yellow flowers with the dark green, rounded, shiny leaves that makes this plant unique. *R. montroseanum* has very good foliage, larger than *R. grande*, yet smaller than *R. sinogrande*: both *R. montroseanum* and the rather similar but smaller-scale *R. pudorosum* have good pink flowers. The leaves of *R. magnificum* have a thin grey indumentum. *R. protistum* (the old *R. giganteum*) is the largest, at up to 30m (100ft). Its glabrous leaves are elliptic to oblanceolate, matt green, with a grey-brown indumentum after many years. The flowers can range from creamy-white and pink to a dark rose, mauve or crimson. As *R. giganteum*, it was awarded an F.C.C. in 1953. The flowers of *R. sidereum* range from a creamy-yellow to a clear yellow or creamy-white to pale pink. It is close to *R. grande*, but has smaller leaves and flowers, and delights by flowering much later. *R. watsonii* is the hardiest member of the subsection, yet perhaps the least impressive in terms of flowers and leaves.

R. sinogrande Balf. f. et W. W. Sm. (Chinese *R. grande*)
tū jiān dùjuān "mucronate rhododendron"

giv

a

p

li

kiii

g

hiii

n

d

ei

gii

n

b

SUBSECTION **IRRORATA** Sleumer

(Series *Irroratum*)

GUAN KAIJUN

*R*hododendron irroratum is called *lù zhū dùjuān* in Chinese. *Lù zhū* in Chinese means "dewdrop", and *dùjuān* means "rhododendron". A dewdrop is a symbol of purity, sincerity, honesty and beauty in China. When people wish to describe the beauty of an object or the honesty of a person, they often use the word *Lù zhū*. It is not surprising that this extraordinary rhododendron species has such a beautiful Chinese name.

Rhododendron irroratum is a very complex species and has an extremely wide geographical distribution range, from Tengchong of Yunnan, China, in the west to Qianxi of Guizhou, China, in the east. Its southern limit of distribution is in Sumatra and the northern limit is in Muli of Sichuan, China. It is therefore not surprising that the species has great variations, especially in its flower colour. In the north of its range, the flowers are white or cream with strong flecks, while in the southern part of its range, the flowers are mauve with only few flecks. In its intermediate distribution area its flower colour is extraordinarily rich – flowers can be white, cream, yellow, orange, pink, red, suffused rose, mauve, purple, and innumerable combinations of these. The species is itself split into four subspecies, the other three being *pogonostylum*, *kontumense* and *ninguenense*.

Rhododendron irroratum is a large shrub or small tree species with a height of 2–9m (6½–30ft). Its leaves are coriaceous, oblanceolate to elliptic and green on both surfaces, the flowers are narrowly bell-shaped. In the wild, it grows in thickets, pine forests or often by itself forms a dense forest at an altitude of 1800–3500m (5900–11,500ft).

The species was discovered in Yunnan in 1886 and introduced to western cultivation around the same year. A very remarkable clone, 'Polka Dot', which has white flowers densely marked with purple dots, received an Award of Merit when exhibited by Exbury in 1957.

The first time I saw *R. irroratum* in bloom in the wild was in the 1970s on a field trip from Yunnan to Xizang (Tibet). I can still remember so well that it was a fine spring day. When I trekked towards the top of the Lion Mountain, which is about 100km (62 miles) north of Kunming in central Yunnan, hundreds of rhododendron trees in full bloom suddenly came into view. When I got close to them, I found they were all *R. irroratum*. I was really surprised by the multicoloured flowers, especially the cream colour and felt attached to this rhododendron ever since.

My most stunning encounter with this rhododendron species was in April 1994 on a trip to Qianxi and Dafang in north-western Guizhou. There, *R. irroratum* forms a dense forest both on its own and mixed with other rhododendron species like *R. delavayi*, covering whole mountain ridges and valleys and spanning hundreds of miles. Standing higher up the mountain it was possible to see endless rhododendron forest with multicoloured flowers stretching in all directions. The scale of the rhododendron forest and the multitude of flower colours were almost beyond belief. I had seen so many fine examples of rhododendron forest during my numerous trips in China, but never before anything like this one in Qianxi and Dafang, where a single rhododendron species could cover hundreds of miles. If I am ever asked to pinpoint the zenith of *R. irroratum*, there is no doubt that my answer would be Qianxi and Dafang in Guizhou, China.

An area of about (512sq. km) 200 square miles has recently been designated as a nature reserve in this region. One particular reason to establish a nature reserve there is to prevent local people cutting rhododendron wood for turnery and for firewood.

R. irroratum Franch. var. 'Polka Dot'
lù zhū dùjuān "dewdrop rhododendron"

a

k*v*

k

k *viii*

k*i*

d

j*i*

b

h*i*

e*i*

h*iii*

h

c*i*

j

Subsection **LANATA** Chamberlain

(Series *Campanulatum*)

Edward G. Millais

*R*hododendron lanatum was discovered by Sir Joseph Hooker in 1851, at Dzongri in Sikkim near the Nepalese border, and south of Mount Kanchenjunga. Here, on wild, rocky terrain at 3600–4200m (12,000–14,000ft), it forms both pure stands on its own, or mixed with *R. campylocarpum, R. wallichii* and, occasionally, *R. fulgens*. It is a very beautiful plant, the upper surfaces of the leaves being covered with a greyish tomentum, which makes it easy to identify at quite a distance. The lower leaf surfaces are covered with a thick coffee-coloured indumentum and the sulphur-yellow flowers, which appear in mid- to late spring, are spotted crimson-red – a striking display.

Unfortunately, coming from an area with a monsoon-type rainfall, it is not an easy plant to grow in the West. There are some fine plants near the west coast of Scotland, where the high humidity and rainfall suit it. *R. lanatum* should be planted in pure humus in the form of leaf mould or peat, and the drainage must be perfect. Here in Surrey I have three plants growing under high pines in almost full shade on a natural, 30cm (12in) bed of pine needles. They do well, provided I remember to keep them well watered, and the sandy soil underneath provides good drainage. Trickle irrigation would be a great help.

Rhododendron flinckii was discovered by R. E. Cooper in 1915, near Bumthang in eastern Bhutan. At the time it was regarded as a form of *R. lanatum* but, owing to the different indumentum, was later named after K. E. Flinck, a Swedish botanist. Until recently, *R. flinckii* was a rare plant and in danger of going out of cultivation. Since 1987, however, there have been several expeditions into eastern Bhutan, and seeds from different locations have been collected. These have shown that this plant is extremely variable – it can be as large as 2.7m (9ft) in height, as in Cooper 2145 (*tsariense magnum* of Davidian), down to 1m (3ft), when it merges with *R. tsariense* (EGM 071 and KR 1755). Like *R. lanatum*, its upper leaf surfaces are covered with a grey tomentum. The main difference is the lower leaf surface, which is covered with a very spectacular, bright rusty-orange indumentum. In windy weather a large plant can change from grey to bright rusty-orange and back every few seconds. There are fine stands of *R. flinckii* growing on the western side of the Rudo La, at 3600–4000m (12,000–13,000ft), and on the eastern side of the summit it grows mixed with many other rhododendrons, mainly *R. bhutanense, R. aeruginosum* and *R. xanthocodon*. The flowers can be either pale yellow or pale pink. The 1987 Rudo La collections are probably all yellow, but flowers were not seen when seed was collected recently on the Sheltang La. Most of the recent collections of *R. flinckii* have rather brighter indumentum than the original collection made by R. E. Cooper. *R. flinckii* is an easy plant to grow, provided it has half-shade and plenty of moisture. I am sure that when it is better known, it will be much sought after, both for its flowers and its foliage.

Rhododendron tsariense was introduced in its pink form from southern Xizang (Tibet) under several numbers in 1936 by Ludlow and Sherriff. Later, in 1938, a yellow form was discovered but is not in cultivation. It is a neat, compact plant, not normally growing to more than 1–1.2m (3–4ft), with much smaller leaves than *R. flinckii* and with flowers that are rich carmine-pink in bud, fading to pale pink or white. It is an ideal plant for a large rock garden or for the front of a rhododendron border.

Rhododendron lanatoides is only just in cultivation – it is represented by one plant in Argyllshire and one in Yorkshire. It was discovered by Ludlow, Sherriff and Elliot in 1947, close to Tongyuk Dzong, north-east of Namche Bawa, in southern Xizang. Unlike other rhododendrons in the *Lanata* subsection, the leaves are lanceolate, with typical lanate indumentum. The flowers are white-flushed-pink. The general appearance is similar to *R. roxieanum*. Unfortunately, until these two plants can be cross-pollinated or perhaps fresh seed collected from Xizang, it will have to remain a collector's dream!

R. lanatum Hooker f. "woolly"

a

b

ci

k

h*iii*

d

e*i*

n

l

k*i*

g

g*ii*

SUBSECTION **MACULIFERA** Sleumer

(Series *Barbatum*, Subseries *Maculiferum*)

KENNETH J. W. LOWES

"*Rhododendron pseudochrysanthum* is one of the finest rhododendrons in cultivation." I quote Mr. Edmund de Rothschild, writing in the 1971 issue of the R.H.S. *Rhododendron and Camellia Year Book*. As it happens, I first saw this species in his garden at Exbury, and I have never forgotten the deep impression it made on me then. Contemplating the overwhelming beauty of the whole plant, I began to reflect that it had come to Exbury from its native habitat without any kind of intervention from gardener or hybridist. Arriving in our gardens direct from its mountain home, *Rhododendron pseudochrysanthum* is now securely in place as one of a select band of favourites in the gardens of rhododendron enthusiasts. We now have a number of slightly different forms of the species – some are dwarfer and quite compact, some are taller, and some are lacking the favoured rich silver-grey coat to the top of the leaf. In the same *Rhododendron Year Book*, in an article concerning their researches into the rhododendrons of Taiwan, J. J. Ravenscroft Patrick and Dr. Chien Chang Hsu say: "With its very short internodes and fine set to the leaves, *R. pseudochrysanthum* vies with the best of dwarf rhododendrons, if not surpassing anything in its size range. Colour slides of dwarf *R. pseudochrysanthum* in flower exhibit all the delights of this magnificent species: in fact, it seems more magnificent than its larger relatives." This was written after a substantial programme of research in the field of the Rhododendrons of Taiwan.

Rhododendron pseudochrysanthum has been placed in Subsection *Maculifera*, which contains a number of similar plants designated as species and having various relevant characters. How many, and which, may not yet be finally decided by the various taxonomists who have attempted their own revisions. The Edinburgh Revision of the early 1980s sets out the following position with nine species: (1) *R. longesquamatum*, (2) *R. ochraceum*, (3) *R. strigillosum*, (4) *R. pachytrichum*, (5) *R. sikangense*, (6) *R. maculiferum*, (7) *R. morii*, (8) *R. pseudochrysanthum*, (9) *R. pachysanthum*. *Rhododendron maculiferum* itself has a subspecies, ssp. *anhweiense*. One or two older names, such as *R. monosematum* and *R. cookeanum*, are at present on the sidelines. Little is known in Britain about *R. ochraceum*, which in the wild can have brilliant red flowers; so too can *R. strigillosum* – but not always, for it can be other colours. A very different natural grouping can be discerned – that of geographical distribution. Whereas *R. morii*, *R. pseudochrysanthum* and *R. pachysanthum* are to be found only outside China – that is, on the island of Taiwan – the other six are native to the Chinese mainland; of these it is only the subspecies *anhweiense* which is to be found quite some distance from Sichuan.

Since 1971 a lot more has been learned about *R. pachysanthum* from Taiwan. This too is proving to be an absolutely outstanding rhododendron for our gardens. So far the plants are relatively low-growing and compact, and the leaves, attractively shaped and very heavily indumented, show that supreme quality which appeals instantly to gardeners, whether connoisseurs or novices. Add to this the chaste beauty of the flower-trusses and a considerable degree of hardiness, and we have a virtually ideal garden plant.

The species *R. ochraceum* and *R. sikangense* are little known and are virtually unobtainable by gardeners. *R. longesquamatum* can be found in good collections; it is perhaps of limited and individual appeal. *R. morii*, *R. pachytrichum* and *R. maculiferum*, usually taller plants from 4–9m (12–30ft), are all attractive in their selected garden forms. Their colour range centres on a white ground with the possibility of various shades of pink, sometimes with blotches, and some spotting, light or heavy. *R. maculiferum* ssp. *anhweiense* is normally compact and dense-growing; it is very hardy and can be exceptionally free-flowering – a great garden success. The last four are all worth a place in the garden, given that there is space and that they are suitably sited. *R. strigillosum* is too early into bloom for a general recommendation, although its colour is commanding. However, unlike the other eight species, it has been successfully used as a parent of spectacular hybrids. Outstanding are 'Lady Digby', 'Matador' and 'Rocket', all rather early-flowering but magnificent in a favourable spring.

To a gardener considering the acquisition of one or more plants of this subsection, I suggest beginning with any of the three compelling, compact growers: *R. pseudochrysanthum*, *R. pachysanthum* and the subspecies *anhweiense*; then, for a larger-growing plant, the very beautiful *R. morii*. I consider all these four to be irresistible, and the plants are all reasonably hardy if located in suitable positions. Their flower trusses are of adequate size to be appreciated as you walk up to them, yet not so bulky or clumsy as to be vulgar. They would look right in a woodland setting. There, they would not disturb the nightingales.

R. pseudochrysanthum Hayata "false *R. chrysanthum*"

a

b

e

h*iii*

k

g*ii*

SUBSECTION **NERIIFLORA** Sleumer

(Series *Neriiflorum*)

PROF. DR. WOLFGANG SPETHMANN

*I*t was the Abbé Delavay who discovered the charming *Rhododendron neriiflorum* above the Tali range in West Yunnan in 1883. It occurs in Xizang (Tibet) and North-East Burma as well, mostly on rocky slopes and rocky meadows, but also in dense pine forests and in mixed forests at elevations of 2300–3600m (7500–11,800ft). In 1910, Forrest sent seeds to England for the first time.

The species grows with a compact and dwarf habit in open places, but is more straggly and higher in shaded places. The deep crimson colour of the flowers gave the plant its name – they look similar to those of *Nerium oleander*. The tubular-campanulate, fleshy corolla with nectar pouches at the base, and the fleshy red-coloured calyx are typical for many species of this subsection. In 1929, the species received an Award of Merit as *R. euchaites* named 'Rosevallon'.

Chamberlain divides the species into three subspecies: ssp. *agetum*, ssp. *neriiflorum* and ssp. *phaedropum*. The subsection *Neriiflora* is divided into four Alliances by Chamberlain: *Neriiflorum, Haematodes, Forrestii* and *Sanguineum*. Besides *R. neriiflorum*, important species of the closely related *Neriiflorum* Alliance for cultivation in our gardens are *R. floccigerum*, *R. sperabile* and *R. sperabiloides*.

The *Haematodes* Alliance is well-known for its deep red flowers and its woolly indumentum. Important species are *R. beanianum*, *R. catacosmum*, *R. coelicum*, *R. haematodes, R. piercei* and *R. pocophorum*. *R. haematodes* is used as a parent for many red or rose hybrids, for example: 'Blitz', 'Humming Bird', 'Little Bert', and the famous 'May Day'.

The *Forrestii* Alliance includes two prostrate, creeping species for our gardens: *R. chamae-thomsonii* and the excellent creeping *R. forrestii*, better known by its old name, *R. repens*. *R. forrestii* is the parent of many of the best, red, elepidote, compact-growing hybrids. The German breeder Hobbie became famous through his cross 'Essex Scarlet' × *R. forrestii* 'Repens' with the worldwide known cultivars 'Baden Baden', 'Scarlet Wonder' and 'Elisabeth Hobbie'.

The *Sanguineum* Alliance is the largest group, with a confusing number of species, subspecies and varieties. Within one species additionally there is often an unusual variation in flower colour, for example *R. dichroanthum* shows rose to carmine, sometimes yellow-flushed, with corollas from yellow to deep orange. Species

for garden cultivation are *R. citriniflorum, R. eudoxum, R. microgynum, R. parmulatum, R. temenium* and, the king in variability, *R. sanguineum*.

Although Subsection *Neriiflora* is a large subsection, the largest of the Subgenus *Hymenanthes*, with 49 species according to H. H. Davidian (1992), but only 26 after the revision of Chamberlain (1982), all the species are found only in a very restricted area: Central East Bhutan, North Assam, North Burma, South-East Xizang and North-West Yunnan – a small band of 150–250 km (95–155 miles) in diameter and with a length of only 800 km (500 miles) from east to west. As a taxonomist this fact leads me to believe that the subsection is phylogenetically young and still developing; new species are separating from others, a large number of subspecies and varieties showing this trend, which is most obvious in the *Sanguineum* Alliance. Another fact supports my view: the former subseries *Sanguineum* is the only unit in the subgenus *Hymenanthes* in which orange flowers occur. That was justification for me to change the status of this group to that of its own subsection: *Sanguina* (Tagg) Spethmann.

In the genus *Rhododendron*, orange colour is only found in the *Vireya* subgenus, Subsection *Cinnabarinum* (subgenus *Rhododendron*) and in the Section *Pentanthera* (subgenus *Nomazalea*). Common to all these taxa is an enormous variability of species, and especially of flower colour. In all these groups, the colour violet is absent; only red, orange, yellow or white occur. Orange colouring is caused by carotenoids only, located in the chromoplasts. Together with anthocyanins, which give the red colours, we get the total spectrum of orange to deep red.

But some may remember also the black-red to brown-red corollas of *R. haemaleum* or *R. didymum*. These colours are mixtures of red and green. Why green? Instead of the previously mentioned chromoplasts, I found chloroplasts (also present in green leaves) in these flowers. Such colouring chromoplasts and chloroplasts I found for the first time in species of the *Sanguineum* subsection, and so this group became a nucleus for new ideas, thoughts and theories.

R. neriiflorum (syn. *R. euchaites*) Franch.

huŏ hóng dùjuān "flame-red rhododendron"

a

k

h*iii*

g

g*ii*

e*i*

f*i*

d

f*ii*

SUBSECTION **PARISHIA** Sleumer

(Series *Irroratum*, Subseries *Parishii*)

MAURICE WILKINS

*T*his subsection is now considered to consist of six species, although other recently described species have been included by Chinese botanists.

Being native to fairly low altitudes in Assam, Burma and Yunnan, these species tend to be a little tender, and are more likely to be seen in the wetter, milder gardens of the west of Britain. Late flowering is another useful characteristic, particularly in west of Scotland collections.

Rhododendron elliottii was named after a friend of Sir George Watt, who first introduced a purple-flowered form from Assam in 1914. The species was reintroduced in 1927 when it was collected in the same location by Frank Kingdon-Ward. A medium- to large-growing shrub, with leaves up to 15cm (6in) long, it is more usually seen bearing brilliant crimson or scarlet, dark-spotted flowers, opening freely in late spring to early summer and into midsummer in cooler places. Late flowering usually means late growth, however, so shelter is necessary to prevent damage to new foliage by early autumn frost.

Rhododendron elliottii has been used frequently to produce late-flowering and somewhat hardier hybrids, notably by Lionel de Rothschild, who considered it one of the best scarlet-flowered species introduced. Brilliant-red 'Fusilier', currant-red 'Kilimanjaro' and blood-red 'Grenadier' are exemplary hybrids.

Rhododendron facetum (meaning "elegant") is a medium to large species from Upper Burma and Yunnan, discovered in 1914 by Frank Kingdon-Ward. Introduced by George Forrest three years later as *R. eriogynum* (now considered synonymous), it flowered for the first time in cultivation at Trewithen in Cornwall, England, in 1926. This species also flowers late, in early and midsummer, with rose to bright red or crimson blooms, and needs moist and mild conditions away from summer heat. The leaves, which can be up to 20cm (8in) long, have a layer of thick, loose indumentum underneath for the first year or so.

Rhododendron facetum has also proved to be a useful parent, producing such well-known hybrids as scarlet 'Tally Ho', raised by J. J. Crosfield of Embley Park, scarlet 'Beau Brummell' and cardinal-red 'Romany Chal', both Exbury hybrids.

Rhododendron huidongense, described by T. L. Ming in 1981, is not yet in cultivation, but is apparently close to *R. elliottii* and *R. facetum*, although generally smaller. It hails from a higher altitude in Sichuan, and may be hardier than others in the subsection.

Rhododendron kyawi was discovered by local collector Maung Kyaw in North-East Upper Burma in 1912, and introduced by George Forrest in 1919 as *R. prophantum*. The species (as *R. agapetum*, now considered synonymous) was seen in flower by Frank Kingdon-Ward in 1926, and he was amazed to see this plant still making quite a show on August 23rd, waxing lyrical over his find – "it wrings from us a shout of delight", he writes in *Plant Hunting on the Edge of the World*, "the ground beneath is red as though strewn with red-hot cinders".

This species is not common in cultivation, being probably more tender than the previous two, but where it does well, in the dampest and mildest places, it is magnificent, with foliage up to 30cm (12in) long, and sticky on both surfaces. The large trusses of flowers, from scarlet to crimson, are produced in mid- and late summer, and new growth is correspondingly late.

Rhododendron parishii is a tender species from Lower Burma, allied to *R. kyawi*, and was discovered around 1882 by the Rev. C. S. Parish, the chaplain at Moulmein, and named after him. It is a large shrub, with broad leaves up to 12cm (4¾in) long, and deep red flowers with darker lines along the lobes. It was described again in 1914 by J. H. Lace, Chief Conservator of Forests in Burma, who rediscovered it in flower in 1912. He may well have introduced the species at about this time, although now it seems lost to cultivation.

Rhododendron schistocalyx is hardier than the other species in the subsection, growing at a higher altitude in western Yunnan, but would need a sheltered position in colder gardens. It resembles *R. facetum*, but produces its rose-crimson or crimson flowers earlier, in mid- and late spring. The species was first found by George Forrest in 1917 and introduced in the same year. The name describes the calyx, which is split down one side.

Two further species are included by Chinese botanists – *R. brevipetiolatum*, which comes from Sichuan, was described by Fang in 1986 and is said to be a smaller version of *R. facetum*, and *R. flavoflorum*, from Yunnan, described by T. L. Ming in 1984, which seems to be the odd man out in the *Parishia* subsection, having pale yellow flowers spotted with red. Neither is in cultivation.

R. facetum Balf. f. et Ward "elegant"
mián máo fáng dùjuān "woolly-ovary rhododendron"

a

gi

li

hiii

d

ji

ci

ei

f

k

n

gii

SUBSECTION **PONTICA** Sleumer

(Series Ponticum)

EDMUND DE ROTHSCHILD

The name "ponticum" in a rhododendron will evoke in most people's minds visions of purple flowers festooning acres of land and being a real nuisance to some as a tangled woodland of no value. *Rhododendron ponticum* originated in the damp beech and alder coppices of the southern sub-Alpine areas of the Caucasus. It was brought to Europe by the Romans and came to England as one of the first rhododendron introductions in 1763. Finding its new home eminently suitable, it soon escaped the confines of the formal garden to establish itself in the surrounding countryside. However, many fine rhododendrons form the *Pontica* subsection.

The species in the *Pontica* subsection are *R. caucasicum, R. smirnowii* and *R. ungernii* from the Caucasus, *R. adenopodum* from eastern Sichuan and western China, *R. chrysanthum* (*R. aureum*) from Siberia, Manchuria and northern Japan, *R. catawbiense, R. maximum* and *R. macrophyllum* from America and Canada, *R. brachycarpum* and *R. fauriei* from northern and central Japan; also from Japan are *R. degronianum, R. makinoi* and *R. metternichii*, and finally from Taiwan and islands near Japan come *R. hyperythrum* and *R. yakushimanum*. The forms from Japan show variation in the indumentum, some being white and others brown.

The species illustrated opposite, which I, along with many others, hold in high regard, is *R. yakushimanum* (from Yaku Shima Island). The "yak", as it is affectionately called by all who are familiar with it, is a very important species. Unique in character, beautiful in flower, attractive in foliage and habit of growth, it instantly inspires its admirers with a desire to grow it. It is tolerant of extreme temperatures, able to endure frosts of -5°C (23°F) and, surprisingly – for its original habitat was in the cool, misty mountains of its island home near Japan – has proved to be quite sun-hardy.

Here in my gardens at Exbury is one of the original *R. yakushimanum* plants, which was sent in 1934 to my father, Lionel de Rothschild. He had requested Koichiro Wada, an eminent nurseryman of Numazushi, Japan, "to supply any plants of unusual character and quality". Among the plants Koichiro Wada selected and sent in response were two *R. yakushimanum*, which were duly planted out in a secluded trial bed. These two plants were extremely small and did not demand much attention, but my father recorded and described them in the *British Rhododendron Year Book*. It was after the war, years later, that the head gardener Francis

Hanger left Exbury to become curator at Wisley, taking with him one of the two plants. This plant had an enthusiastic reception at the Royal Horticultural Society's Show in 1947, where it was awarded the prestigious First Class Certificate (F.C.C.). Its companion was exhibited three years later and was met with equal enthusiasm. This very plant can still be seen in my garden: now in its sixth decade and showing signs of age, it nevertheless still produces a yearly display of magnificent blooms.

In its best form very few plants of many species can compare with the ethereal, porcelain-like beauty of the *R. yakushimanum* flowers: these emerge from softly-felted, tightly-packed, ball-shaped buds, gradually unfurling into dome-shaped clusters of 10–12 wide, bell-shaped flowers of a delicate shell-pink hue. As the truss expands to full openness, it becomes a pure porcelain-white, with only a fine sprinkling of green spots freckling the inside. The trusses sit closely in the leather-like dark green leaves, a perfect foil for such beauty.

An added bonus for the gardener is a repetition of this contrast a few weeks later, when the new leaf growth appears: this is covered with a silvery-white, felt-like indumentum, which gives the feel of young rabbits' ears, and indeed resembles them – the new leaves stand upright from the mature foliage and the effect is very ornamental. Gradually, this new growth turns to leathery, dark green leaves about 10cm (4in) long, recurving at the edges and giving a long, spoon-like appearance. They are heavily covered below with a cinnamon-coloured indumentum. In habit *R. yakushimanum* is almost dome-shaped, growing low and compact and taking years to attain a height of 1.2m (4ft).

This comparatively new arrival has made an enormous impact in the hybridization world, enabling the horticulturist to develop an impressive range of low-growing rhododendron hybrids that suit today's suburban garden with its limited space. Among its progeny are many excellent, award-winning hybrids that have inherited the yak's propensity for abundant flowering, for example the series aptly named after the dwarves in *Snow-White*. This includes *R.* 'Happy' (Award of Merit [A.M.] 1977). Then there is the exceptional *R.* 'Seven Stars' (F.C.C. 1974) and the quite exceptional *R.* 'Wiseman' (A.M. 1982). The list is endless – and so is the pleasure.

R. yakushimanum Nakai (from Yaku Shima Island, Japan)

a

k

b

j

li

kii

d

ei

hiii

g

gii

ci

SUBSECTION **SELENSIA** Sleumer

(Series Thomsonii, Subseries Selense)

RICK PETERSEN

One of the foremost collectors of species rhododendrons was the Scotsman George Forrest. He spent 28 years exploring for plants in the Chinese provinces Sichuan and Yunnan. During the time he was in China, Forrest gathered seed from more than 5000 rhododendron plants and discovered over 300 new species of rhododendron. Indeed, he found the majority of species in Subsection *Selensia*.

In the most recent revision of the genus *Rhododendron*, Subsection *Selensia* now includes seven species: *R. bainbridgeanum*, *R. calvescens*, *R. dasycladoides*, *R. esetulosum*, *R. hirtipes*, *R. martinianum* and *R. selense*.

Originally placed in the *Barbatum* series, *R. bainbridgeanum* was moved to the *Selensia* subsection because of its closer relationship to the species in that group. In 1922, Forrest discovered this species in south-eastern Xizang (Tibet), and its range extends into north-western Yunnan and adjacent areas of northern Myanmar (Burma). The plants have flowers that are white or creamy-white, often flushed pink or rose, pale to deep pink or creamy yellow, and may have flecks and a blotch.

Rhododendron calvescens was found by Forrest in 1917 on the Mekong-Salween Divide in south-eastern Xizang and it is also native to north-western Yunnan. The flowers may be white, white-flushed-rose or rose, and may or may not have crimson flecks. The species is divided into two varieties, var. *calvescens* and var. *duseimatum*.

First collected by Camillo Schneider in 1914, *R. dasycladoides* is still not in cultivation. This species is found in the forests of south-western Sichuan and north-western Yunnan. The flowers may be pale pink to rose or pink-purple with crimson flecks.

Forrest also discovered *R. esetulosum* in the mountains of north-western Yunnan in 1918, and the range of this species extends into south-eastern Xizang as well. Plants may have creamy-white or white flowers that are sometimes flushed rose or purple, with or without crimson or purple flecks.

Another species that was transferred from the former *Barbatum* series is *R. hirtipes*, which was first found by Frank Kingdon-Ward in 1924. The distribution of this species is south-eastern Xizang, in an area some distance to the west of the other members of this subsection. The campanulate flowers may be almost white to white-flushed-pink or pink, with or without pink striping, carmine, purple flecks or purple blotch.

Possibly the most captivating species within this subsection is *R. martinianum*. Forrest found this attractive rhododendron growing in north-western Yunnan in 1914 and it is also native to south-eastern Xizang and north Burma. Flower colour ranges from white or creamy-white, sometimes flushed rose, to pink, rose, or pale yellow, and often the inner petals are marked with crimson or purple flecks.

The central species of this subsection is *R. selense*, which is a complex and variable mixture of subspecies. They are gregarious plants that hybridize with each other in the wild and with other rhododendrons, making classification difficult. Reclassification has made a number of former species synonymous under *R. selense* as well as submerging several as subspecies. Forrest discovered three of the four subspecies: ssp. *dasycladum* in 1913, ssp. *jucundum* in 1906, and ssp. *setiferum* in 1917. The fourth, ssp. *selense*, was first found by Abbé J. A. Soulié in 1895.

Rhododendron selense ssp. *dasycladum* is native to north-western Yunnan and south-eastern Xizang. Blossoms may be white to pink or rose.

The most distinctive feature of ssp. *jucundum* is the glaucous bloom, that is, a blue-green waxy coating, on the undersides of the leaves. This subspecies is found in north-western Yunnan, some distance to the south of its relations. The flowers are white to pink or rose.

Rhododendron selense ssp. *selense* is found in north-western Yunnan, south-western Sichuan and south-eastern Tibet. The flowers are white to pink or rose.

Although it has been given taxonomic standing, ssp. *setiferum* is felt by many to be a natural hybrid, possibly between *R. bainbridgeanum* and ssp. *selense*. These rhododendrons range from north-western Yunnan into south-eastern Xizang. The flowers are creamy-white.

In the area of Asia where these rhododendrons are native, there is still a great deal of research and collection that needs to be done to sort out the relationships of these plants and to bring into cultivation new forms for our gardens.

R. selense ssp. *jucundum* Balf. f. et. W. W. Sm. "pleasant"
duō biàn dùjuān "variable rhododendron"

a

b

g

gvi

d

ei

ci

oi

gii

hi

h

hv

k

ki

SUBSECTION **TALIENSIA** Sleumer

(Series *Taliense*, Series *Lacteum*)

WARREN BERG

The *Taliensia* subsection is a large, complex group of plants with uncertain relationships. With the addition of the former *Lacteum* series and 13 new species, it is even more obscured, making a total of more than 60 species. Horticulturally speaking, these species include some of the most beautiful in the genus. Many have outstanding foliage, with variations of indumentum and tomentum to enhance their charm, as well as many other fine qualities, including hardiness, form and often compactness. They are usually long-lived and are quite free-flowering when mature. In the past, this subsection was rather looked down upon, especially by Kingdon-Ward, because they were so slow to bloom and rather difficult to root. However, with improved propagation methods and a greater tolerance for the time it takes to flower, their popularity in the garden has grown considerably in recent years.

Of the 13 new *Taliensia* that Peter Cox mentions in *The Larger Rhododendron Species*, I have been fortunate enough to find several in my explorations. These include *Rhododendron bhutanense*, which was found in Bhutan on the top of the Rudong La at 4150m (13,600ft). This is a very fine new species, and is probably the most westerly representative of this subsection. While in the upper Wolong Panda Reserve in Sichuan in 1983, I found *R. balangense*. It is a very handsome plant with white flowers and white indumentum, growing at nearly 3000m (10,000ft). As Peter mentions in his book, the classification may be a little suspect, but the Chinese consider it to be a *Taliensia*. Another new one which happens to be particularly attractive is *R. coeloneurum*. We found it on two different expeditions on the east side of the Erlang Shan in South-East Sichuan, at about 2800m (9300ft). It is closely related to *R. wiltonii*, but has more bullate leaves and a much thicker indumentum. It may be the most easterly *Taliensia* and possibly the least hardy because of the elevation at which it grows; however, it is certainly well worth having.

It is very difficult to pick the finest member of this very interesting subsection, but for me it probably comes down to *R. bureavii* or *R. proteoides*. Since my years of searching for the latter in Yunnan, Sichuan and Xizang (Tibet) have been to no avail, I have chosen the former as the best overall representative. I have found it both in Yunnan and Sichuan. I believe there would be very little disagreement that a properly grown, mature *R. bureavii* in its best form, such as the Exbury Award of Merit, is about as handsome as any species in the genus. It is normally easy to grow, when given partial shade. The dark, shiny, bullate foliage with the rich, rusty-red indumentum is absolutely outstanding, even for the most discriminating. When mature, it is free-flowering with flushed-rose or white-suffused-rose blooms, fading to white and most often with crimson spots which enhance its beauty. There are usually 10–15 flowers in the truss.

Rhododendron bureavii is found both in Yunnan and Sichuan at altitudes from as low as 2700m (9000ft) up to over 4400m (14,500ft). The Chinese consider the plants in Sichuan to be *R. bureavioides*, but these have been merged into *R. bureavii*. The difference is not significant. In 1983 I found a large population of *R. bureavioides* in North-West Sichuan near Mount Sigunion. These were growing only along the rivers and streams at about 3660m (12,000ft), with many fine forms having a heavy indumentum. Some of the bushes were as high as 7.6m (25ft). This trip was made in autumn, so I am unable to report on the flower. On another expedition in 1989, just north of Kanding in western Sichuan, we found another fine population. It was in a forest preserve called Chi-Sa-Hi. Again the *R. bureavii* were growing mostly along the river and lake shore and because this was a reserve, many of the plants were very large. One in particular, the best form I have seen in the wild, was at least 6m (20ft) in diameter and about 7.6m (25ft) high. It was very heavily indumentumed, with a layer of tomentum on the upper surface of the leaves, and altogether an impressive plant.

It is only in recent years that *R. bureavii* has been used much in hybridization. Halfdan Lem was one of the first in the United States to use it in his crosses. This was with *R.* 'Fabia', of which he named one 'Hansel' and the other 'Gretel'. These hybrids, and many since by other hybridizers, all have beautiful foliage and habit, inherited from *R. bureavii*. I used it with *R. yakushimanum* in 1965, and others have also made this cross, all with good results, although I must admit that some are slow to bloom.

As perhaps you have surmised from the above, *R. bureavii* is probably my favourite species and is therefore well represented in my garden.

R. bureavii Franch. (after E. Bureau 1830–1918, a French professor)

SUBSECTION **THOMSONIA** Sleumer

(Series *Thomsonii*)

DAVID GOHEEN

Of the many plants introduced to cultivation by Sir Joseph Hooker, from the mountainous regions to the north and east of India, few are more respected and admired than the blood-red *Rhododendron thomsonii*. This plant has been rated by generations of gardeners as being in the first rank of garden ornamentals. Its foliage and flowers, combined with its adaptability to diverse garden sites and situations, have caused it to be considered as one of the very best species in the genus *Rhododendron*. Hooker found the plant in 1849 in the Himalayas, sent seed to England and named it in honour of his travelling companion, Dr. Thomas Thomson.

Rhododendron thomsonii is the central species of a fairly numerous assemblage of casually related species which may be found along the mountainous regions from Nepal, Sikkim, Bhutan and into Xizang (Tibet). As one goes further east into Upper Burma and on into Yunnan Province in China, *R. thomsonii* merges into a separate species, *R. meddianum*, often called the Chinese *R. thomsonii*. It can be distinguished from the Himalayan *thomsonii* by its shorter stature, mature specimens being less than 2.5m (8ft) tall, its non-glabrous ovary and its ovate, egg-shaped leaves. In other aspects, it is much like *R. thomsonii* — both species have fleshy, scarlet-red flowers that in cool situations may last as long as four weeks.

At one time, in the Balfourian system of rhododendron classification, a fairly large number of species were considered to be closely related to *R. thomsonii* and were placed in what was termed the *Thomsonii* series. Many of these species have now been reclassified in the Sleumer, Cullen and Chamberlain system of dividing the genus *Rhododendron* into sections and subsections. Thus a smaller number of species are now considered as being closely related to *R. thomsonii* in Subsection *Thomsonia*. Several very beautiful species are included in this subsection, among which are the previously mentioned *R. meddianum*, *R. hookeri*, *R. stewartianum*, *R. eclecteum* and the very beautiful *R. cerasinum*. Kingdon-Ward considered the latter species especially fine and named one of its clones in the field 'Cherry Brandy'.

Rhododendron thomsonii is generally found as a shrub or small tree up to about 6m (20ft) in height. I have seen specimens in the tree category in Scotland, grown from seed sent back by Hooker, that are glorious sights covered with masses of deep red flowers in trusses of 6–12 florets. It is one species that probably should be fertilized every year, because it has a tendency to produce multitudes of flowers and to virtually exhaust itself by its gorgeous display. The individual bell-shaped, campanulate florets are 5-lobed and are deep blood-red in colour. They may be 5cm (2in) long and up to 7.5cm (3in) wide in the best forms of this variable species. Of particular interest is the large calyx, which is cup-shaped, as long as 2cm (¾in) and varies from red to green.

Not only does *R. thomsonii* shine at flowering time, it also presents an appealing appearance in the garden all year with its attractive, smooth, peeling bark and its new emerald-green growth with a fleeting waxy bloom that changes to a blue haze as it matures. These characteristics, combined with its rounded, glabrous leaves, up to 10cm (4in) long and 5cm (2in) wide, blue-green above and whitish below, make this species an outstanding plant.

Altogether, *R. thomsonii* has very few adverse traits. One can fault it with only a few minor criticisms: it does seem to take a longish time to come into flower when grown from either seeds or cuttings, but it makes up for the wait by blooming heavily and reliably after the first blooms appear. One other concern is the fact that it does not respond well to pruning. If branches are cut, new sprouts from older wood are rarely produced. Thus, shaping needs to be accomplished by judicious pinching of new growths.

Rhododendron thomsonii has been used rather extensively in hybridization, especially in early efforts to obtain good, clear reds. Many of its hybrids have been given awards. Among these are 'Sir John Ramsden', 'Aries', 'Barclayii Robert Fox', 'Exbury Cornish Cross' and 'Bodnant Thomwilliams'. Many other *thomsonii* hybrids are well known in horticultural circles. *R. thomsonii*, even after being in cultivation for almost 140 years, continues to bring charm and beauty to our gardens and it is still in hybridization even after the passage of many years.

In my own garden, I await the first blooms from a plant grown from a cutting from a fine *R. thomsonii* from the Crystal Springs Rhododendron Garden in Portland, Oregon. Although my plant has yet to flower, I have seen the parent plant in its full glory and I look forward eagerly, yet patiently, to seeing my plant when, in the fullness of time, it decides to flower!

R. thomsonii Hooker f. (after Thomas Thomson, 1817–1878, a former Supt., Calcutta Botanic Gardens)
bān tuán yè dùjuān "semicircular-leaved rhododendron"

a

c*i*

b

e*i*

d

h*iii*

g

k

g*i*

SUBSECTION **VENATORA** Chamberlain

(Series *Irroratum*, Subseries *Parishii*)

PETER CLOUGH

*R*hododendron venator has to me and others always seemed a little of a misfit. I grew it first in a woodland glade at Achamore Gardens on the Isle of Gigha, Scotland. I knew it as a rather open, 1.8m (6ft) high, "see-through" shrub, with not unattractive, rough-surfaced, medium-sized leaves. Sir James Horlick, my rhododendron teacher, showed me the taxonomist's problem. The young shoots bear long fine white hairs mixed with glandular bristles. The botanists, unconvinced, had at that time filed it into the *Parishii* subsection of the *Irroratum* series. To a simple gardener (if there is such a thing) it looked as though it bore closer relationships to *R. barbatum*. The latest classification accepts its singularity by placing it alone in its own specially created subsection, *Venatora*.

Its value in the woodland garden is found in the flower. At its best, planted in a sufficiently open situation to reduce its tendency to "gawkiness", its flower trusses can be startling, composed of a bright, loose assembly of scarlet-orange tubular bells, enhanced by deep, alluring nectar pouches.

The name "venator", meaning "the hunter", refers to the flower colour – the scarlet-orange of the huntsman's jacket, ever-known in our curious English idiom as "Hunting Pink". It could be added that *R. venator* bears other characteristics reminiscent of the "hunting set" in that it is pretty hardy and distinctive with a tendency to be rather "leggy".

Rhododendron venator has not often been visited or collected in the wild. It has a limited distribution deep in the Tsangpo gorge in South-East Xizang (formerly Tibet), where it is found in scrubby, swampy thickets and clinging to slimy rock faces. At the moment these areas are difficult to approach both politically and, even more, practically, due to the poor state of the road systems in that area. It may be that *R. venator* will remain aloof there for some time hence.

The Bodnant Garden form was given an Award of Merit in 1933, which may have sparked the interest of a few hybridizers with the lateness of its flowering, often into early summer. Most of its hybrids seem to date from the 1940s. Many are now lost or have been superseded. Exbury produced 'Isme' and 'Mandalay'. Bodnant gave us 'Vanven', 'Venco' and 'Veta' (the parents are 'Vanessa' and *R. venator*, 'Coreta' and *R. venator* and 'Etna' and *R. venator* respectively). Thacker produced 'Knowle Brilliant', and 'Fire Music' with *R. dichroanthum* as the other parent. Probably the best-known of the surviving cultivars are Headfort's 'Vanguard', a scarlet *R. griersonianum*-hybrid; 'Grayswood Pink' with strong *R. williamsianum* blood, and 'Venapens' produced at Muncaster by Ramsden, derived from *R. forrestii* var. *repens*, and still well worth choosing as a compact scarlet.

R. venator Tagg. "the hunter" (alluding to the scarlet colour of the flower)

a

li

j

oi   ei   h

b   ji   ci   g   k

d

SUBSECTION WILLIAMSIANA (Cowan & Davidian) Chamberlain

(Series *Thomsonii*, Subseries *Williamsianum*)

JULIAN WILLIAMS

If one was asked to name an example of a rhododendron that had changed the range of plants available to gardeners around the world as a result of the struggles and privations suffered by the great British plant hunters in the first third of the twentieth century, then Subsection *Williamsiana* and within it the *R. williamsianum* are good plants to be considered.

R. williamsianum is a permanent reminder of what is owed to the great Chinese plant collector E. H. Wilson and to John Charles Williams of Caerhays Castle. Introduced to cultivation by E. H. Wilson when he started to work for Veitch's Nursery between 1900 and 1904, the rhododendron seedlings were first tried and tested in John Charles Williams' garden in Cornwall. The success of the cultivation of Chinese seed in Cornwall encouraged Wilson to return to China, this time working for Professor Sargent and the Arnold Arboretum.

Wilson, in his *Plantae Wilsonae*, describes *R. williamsianum* (No. 1350) as "a pretty and distinct species – similar to but different from *Rho. orbiculare* which was thought to be a much more vigorous species. This new species is apparently very local occurring only in isolated places in the cliffs of Wa Shan". He then continues: "…this species is named for Mr. J. C. Williams of Caerhays Castle, Cornwall, England. The first amateur to appreciate the horticultural value of the rhododendrons of Western China; in his garden the best collection of these new varieties is now to be found." (*Plantae Wilsonae*, Vol. 1, p.538)

The plant hunters were very proud of – and fearful for – the future of the plants they introduced. Somewhere in Caerhays' records there is a letter from Wilson to J. C. Williams which accompanied the six plants sent to Caerhays in 1912 from the Arnold Arboretum. In this he states that great care should be given to their cultivation, as he had only found the rhododendron in one place, which had been an extremely difficult site to reach. He doubted whether he would ever be able to return to the place, and if he ever did, it was extremely likely that the plant would have been destroyed by fire or some other disaster.

So Wilson's consignment of plants brought with them great responsibility. This J. C. Williams accepted, but, as a precaution, he sent two plants to his cousin P. D. Williams at Lanarth on the Lizard Peninsula. Fortunately, *R. williamsianum* comes easily from cuttings, and it is likely that from these six plants came a high percentage of the plants growing today in European gardens.

Wilson's admonitions were heartfelt. If one looks at the photographs he took of Mount Wa Shan in 1908, one can see what a wild and terrifying place it is. There are photographs showing sheer drops of 610m (2000ft) to rivers flowing furiously below. Mount Wa Shan, 3400m (11,210ft) high, rises steeply and looks far from hospitable. As we enjoy the fruits of the labours of men like Wilson, Forrest, Farrer and Kingdon-Ward, the hardships and dangers they endured should never be forgotten.

It was in this wild region that *R. williamsianum* was found in February 1908. The rhododendrons reached Caerhays in the spring of 1912, sent from the Arnold Arboretum. On 3 May, 1912, E. H. Wilson wrote to J. C. Williams: "I am very glad the little plants of 1350 reached you safely. It will interest you to learn that I have since located another small stock of this number. Next Autumn I will endeavour to send you a couple more plants. No. 1350 is a new species most closely allied to *Rho. Souliei* and *Rho. Orbiculare*. It is very distinct and is not closely related to any known species."

Rhododendron williamsianum, with its pink flowers and dome shape, gradually acclimatized to Caerhays and, although the shrubs had reached only 30cm (1ft) in height by 1922, they had become substantial bushes of 1.8m (6ft) or so 50 years later. The original plants have now all died. It appeared to the writer that their leaves were so closely packed together that it was difficult for the plants to feed themselves properly. Other rhododendron species grown from original seed still survive at Caerhays 90 years on.

Although *R. williamsianum* is a slow grower, it has been a very successful parent, producing many fine rhododendron hybrids. Britain has produced *R.* 'Pink Pebble', *R.* 'Thomwilliams', *R.* 'Humming Bird', *R.* 'Cowslip' and *R.* 'Moonstone'; Europe has produced *R.* 'Gartendirektor Glocker', *R.* 'Moerheim's Pink' and *R.* 'Osmar'; the United States *R.* 'Kimberly' and *R.* 'Kimbeth'.

From the gorges and ravines of Wa Shan in Sichuan, China, *R. williamsianum* has travelled a long way and, whether as itself or as a parent for hybrids, it has given a great deal of pleasure to thousands who know not of its origin nor its travels.

R. williamsianum Rehd. et Wils. (after J. C. Williams, 1861–1939, of Caerhays, Cornwall)
yuán yè dùjuān "round-leaved rhododendron"

a

b

d

gii

hiii

ei

eii

g

k

PART II
SUBGENUS RHODODENDRON

*T*here is no doubt in my mind that having to represent the whole of the genus *Rhododendron* in this book forced me prematurely into discovering this subgenus with its three sections. I dare say it would have been a pleasure to have continued painting the *Hymenanthes* rhododendrons and, with over 220 species to select from, that would have occupied me for many more seasons. But how much I would have missed! On first appearance, so many of the tiny-flowering species, some with the smallest of leaves, didn't seem so exciting, especially as the *Hymenanthes* beauties kept beckoning and tempting me to return to them. But I resisted and was rewarded. Today I can appreciate fully how many collectors of species have decided that this is the subgenus for them. The old adage that "good things come in little packages" has great relevance when considering this subgenus.

To the north of the Gardens at Exbury are the main plantings of species that belong to section *Rhododendron*. Dense thickets of *R. cinnabarinum*, with their glowing fiery colours and distinctive, tubular, bell-shaped flowers can be seen here. I remember standing in the middle of a group of these 6m (20ft) high plants late one spring evening. The setting sun was forecasting another fine 'shepherd's delight'-day by sending golden-red shafts of light flickering through the silhouetted branches and leaves and the waxy flowers of *R. cinnabarinum* which were lit up with such a vibrancy of jewel-like colours of rubies, beryls and garnets, that I remained transfixed by their beauty. A few yards west of this delight is Augustinii Corner which, like the *R. cinnabarinum*, are the original plantings of Lionel de Rothschild's *R. augustinii*. Here the opposite effect can be seen – the cool-blue, lavender-mauve colours of this species with their delicate, butterfly-shaped flowers, shimmer and shake at the slightest ruffle of a breeze. They bloom with great abundance and, as they have reached heights of 6m (20ft) or more, prove to be a popular attraction to the visitors.

The strongly-scented species are to be found in this subgenus. Many species in subsection *Maddenia* have lily-like flowers and perfumes to match. *R. lindleyi* and *R. crassum* were overwhelming in the confines of the studio, although to savour their scents carried on the breeze from the Solent Water in the Gardens is a delight. *R. edgworthii's* perfume had been confined to an inflated plastic bag and box on its journey from the Royal Botanic Garden

at Edinburgh; when opened, it nearly sent me reeling across the room its perfume was so potent.

Along Lover's Lane at Exbury are small tree-like plants of *R. racemosum*, a species that could be grown as bonsai, their twisting habit and small blossoms making them ideal for this technique. The main plantings for these and other small species of this subgenus, however, can be seen in the exceptional two-acre Rock Garden, probably one of the largest of its kind in Europe. Considerable effort has been made in these past few years to enhance the original plantings with extra species, rock plants and spring flowers. This is where *R. lapponicum*, *R. impeditum*, *R. campylogynum*, *R. saluenense* and other ground-hugging species can be found, while mature plants of *R. yunnanense* and *R. oreotrephes*, mixed with *Pieris* and other woodland plants, fringe the skyline.

Part II also holds the species from the second section in this subgenus, the tiny, beautifully-formed flowers of section *Pogonanthum*. These are the aromatic plants which Peter Cox describes so beautifully in his article and his books. They are also the most demanding plants to paint, with their minute flowers and leaves presenting a great challenge.

The third section contains the subtropical rhododendron species, the *Vireyas*, known generally as the 'Malaysian species'. These are very tender, mainly epiphytic species originating from South-East Asia, New Guinea, the Phillipines and other areas. In this country, they can only be grown in conservatories or specialist temperate houses, such as the ones at Kew Gardens or the Royal Botanic Garden, Edinburgh, where Dr. Argent is preparing his revision of the classification of some 200 species in seven subsections which make up this section of spectacular, jewel-like plants.

The watercolours for Part II Subgenus *Rhododendron* were painted over a four-year period from species collected at Exbury, with the exception of: *R. afghanicum*, *R. baileyi*, *R. minus*, *R. edgworthii*, *R. micranthum*, *R. ferrugineum*, *R. fragariflorum*, *R. ledum* and *R. monanthum* which were supplied by the Royal Botanic Garden, Edinburgh; *R. christii* from the Temperate House, Kew; and *R. sulfureum* and, *R. camelliiflorum* courtesy of Mr. Peter Cox, Glendoick, Scotland.

Section *Afghanica*, see page 104.

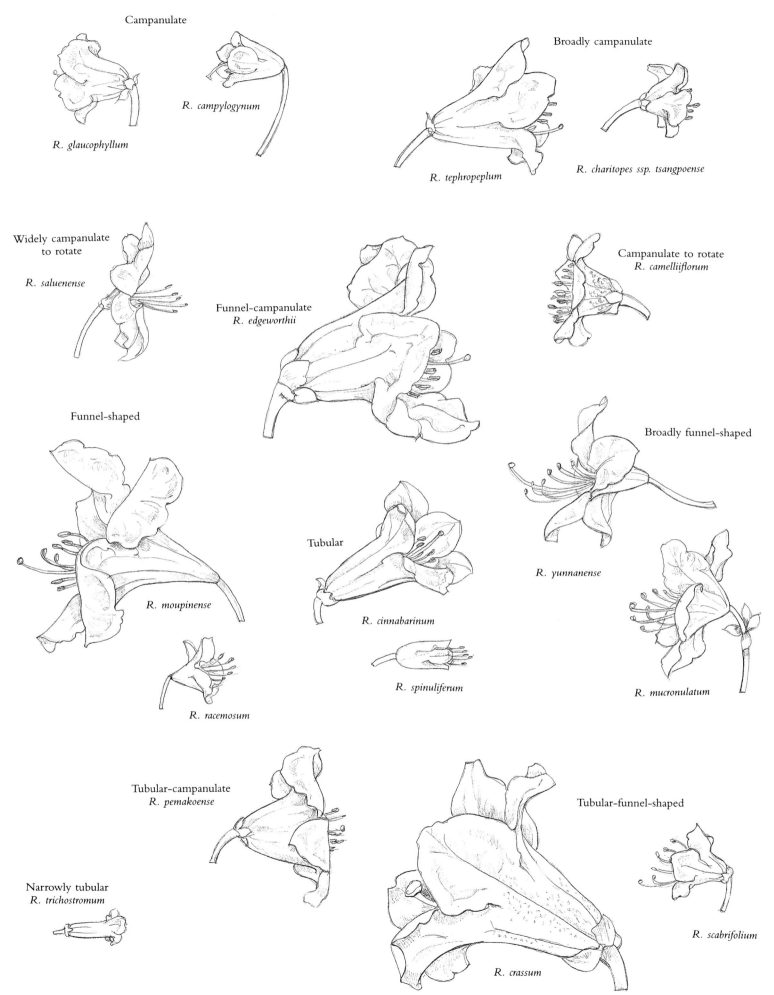

Campanulate

R. glaucophyllum

R. campylogynum

Broadly campanulate

R. tephropeplum

R. charitopes ssp. tsangpoense

Widely campanulate
to rotate

R. saluenense

Funnel-campanulate
R. edgeworthii

Campanulate to rotate
R. camelliiflorum

Funnel-shaped

R. moupinense

Tubular

R. cinnabarinum

R. spinuliferum

Broadly funnel-shaped

R. yunnanense

R. mucronulatum

R. racemosum

Tubular–campanulate
R. pemakoense

Tubular–funnel-shaped

Narrowly tubular
R. trichostromum

R. scabrifolium

R. crassum

FLOWER SHAPES OF LEPIDOTE RHODODENDRONS

SUBSECTION **BAILEYA** Sleumer

(Series *Lepidotum*, Subseries *Baileyi*)

BRUCE ARCHIBOLD

*R*hododendron baileyi is the only species in this subsection and was named after Lieutenant Colonel Bailey, who collected it in Xizang (formerly Tibet). As well as being found in the South-East corner of that country, it also grows in Sikkim and Bhutan, where it is found at heights between 2400 and 4000m (8000 and 13,000ft), so being reputedly not quite hardy. However, it grows and flowers quite happily in some of the colder parts of the United Kingdom, provided it can be given some shelter.

Although inclined to be leggy, it is well worth a place in any garden for the flowers, which are of a brilliant purplish-red colour and held high on long pedicels, rather like a periscope. If the plant can be placed in a spot where the late sun can shine through the flowers, a very beautiful effect can be obtained. As a maximum height of around 1.8m (6ft) is usually reached, it is well suited to the smaller garden.

It does not appear that *R. baileyi* has been used for hybridizing. All in all *R. baileyi* is a rather under-appreciated plant which deserves to be more widely grown.

Opposite above: *R. baileyi* Balf. f (after Lt.-Col. F. M. Bailey, 1882–1967, traveller in Tibet)
fú huä dùjuān "wheel-flower rhododendron"

SUBSECTION **BOOTHIA** (Hutch.) Sleumer

(*Boothii* Series)

BRUCE ARCHIBOLD

*O*f the species making up the *Boothia* subseries there are three that are most likely to be met: *Rhododendron chrysodoron*, which is rather romantically translated as 'Golden Gift', since a plant of this yellow-flowered species was given to the Royal Botanic Garden, Edinburgh, by the Earl of Stair; *R. sulfureum*, so named for the sulphur-yellow colour of its flowers; and finally *R. leucaspis*, or 'White Shield', which has milky-white flowers with which the dark brown or black stamens form a pleasant contrast.

All three are found growing on cliffs or rock faces at comparatively low levels, between 2000 and 3660m (6500 and 12,000ft), in the area Yunnan, Upper Burma and South-East Xizang (formerly Tibet), so they are not entirely hardy. *R. leucaspis* is the hardiest of the three, although the flowers are very susceptible to early frosts, being produced in early and mid-spring. The other two are usually found in the kinder climate of the west side of the United Kingdom; but recent introductions of *R. sulfureum* from more exposed locations give hope that hardier plants will be made available in due course. The Award of Merit has been given to all the species mentioned with *R. leucaspis* receiving, in addition, a First Class Certificate.

I must confess to a liking for yellow-flowered rhododendrons, so my choice is reduced to two — *R. chrysodoron* and *R. sulfureum* – both of which, in their best forms, produce excellent deep yellow flowers. Ever since I came upon *R. sulfureum* in Peter Cox's excellent wild garden at Baravalla, I have admired it and now have it, grown from seed obtained from that other fine garden, Brodick Castle on the Isle of Arran, Scotland.

Some good hybrids have come from the three species, for example *R.* 'Chrysomanicum' (*R. chrysodoron* × *R. burmanicum*), *R.* 'R. W. Rye' (*R. chrysodoron* × *R. johnstoneanum*), *R.* 'Ptarmigan' (*R. leucaspis* × *R. microleucum*), *R.* 'Bric-a-Brac' (*R. leucaspis* × *R. moupinense*), *R.* 'Yellow Hammer' (*R. sulfureum* × *R. flavidum*), *R.* 'Busaco' (*R. sulfureum* × *R. moupinense*).

All the hybrids mentioned have been awarded an Award of Garden Merit by the Royal Horticultural Society. This award is given only to an excellent garden plant which has had to undergo a period of assessment under garden conditions.

Opposite below: *R. sulfureum* Franch
liú huáng dùjuān "sulphur rhododendron"

kiii

a

kvii

ci

kv

b

d

mi

ei

gii

gv

c

k

hiii

b

d

hiii

f

j

gii

jii

ei

SUBSECTION **CAMELLIIFLORA** (Hutch.) Sleumer

(Series *Camelliiflorum*)

KEITH RUSHFORTH

*R*hododendron camelliiflorum in name evokes the majesty of both rhododendron and camellia, implying to those who have not seen it a munificence of beauty. Some would argue that anticipation is very often much better than the actual event. On a macroscopic level, this is certainly the case with *R. camelliiflorum*. From a distance, it is scarcely inspiring and therefore it is difficult to argue with Bean's comment, "quite uncommon (in cultivation), which is no matter for great regret, for it is one of the least ornamental and most difficult rhododendrons". However, it is there and is rather unique, occupying its own subsection, and thus commandeering its space in this book.

Rhododendron camelliiflorum was first reported by Hooker from Sikkim and introduced in 1851. It is also recorded from East Nepal and from throughout Bhutan. Further east it extends into Arunachal Pradesh and adjacent areas of Xizang (Tibet), although quite how far east is not proven.

It is in Bhutan that I am familiar with it. There it is common in moist localities from about 2500m (8200ft) up to 3540m (11,600ft). However, it is less often noticed. It appears to be more or less habitually an epiphyte, carrying the genre's invisibility to new peaks of perfection − it neither flowers on a scale to draw attention to itself (such as epiphytic *R. lindleyi*) nor attracts by assuming brilliant red autumn colours (which is what *Sorbus wallichii*, probably the commonest woody species in moist oak forests at around 2700m/8850ft, suddenly does). In all its characteristics it is discreet.

The plant that provided the material for the illustration was collected on the Dochu La (also spelt Dochong or Dokyong La), between Thimphu and Wangdi Phodrang in western Bhutan. It was found at 2835m (9300ft) in the midst of the *R. lindleyi* range, with *Quercus oxyodon* and others. It is more common (or perhaps I've just noticed more of it?) growing on thick crusts of moss in the tops of

Tsuga dumosa on the crest of the La at 3100m (10,160ft), above *R. kesangiae, R. falconeri, R. arboreum, R. keysii* and *R. barbatum*. It is also found on the much drier western side, growing on *Quercus semecarpifolia*, where sufficient moisture-laden air reaches it. Further east I have seen it on the Pele La, the Ura-Sheltang La and on the eastern flank of the Thrumseng La.

This rhododendron is variable, especially in the width of its leaves. Gradually, these fall into two groups: either narrow elliptic or more balanced. However, the plants on the Thrumseng La have leaves that are very broad in relation to their length. They are always marked by an indentation along the midrib. The carpel number is also extremely variable, from 5–10, and appears to have no taxonomic significance. Its short, bent style shows the plant's relationship to the *Boothia* subsection.

The normal flower colour is described as white to rose. However, as is obvious here, the Dochu La plant has a much superior bloom. To quote from *A Quest of Flowers*, when Ludlow and Sherriff first encountered this rhododendron in 1933: "Two rhododendrons fascinated them chiefly because the flowers were so strongly reminiscent of those of other genera. One, with correalike, narrow tubular flowers, was *R. keysii*. The other, with small camellia-like flowers, was *R. camelliiflorum*, and a splendid form of it with deep wine-red fleshy flowers, as dark in colour in fact and just as fleshy, as the port wine corollas of *Buddleia colvillei* which grew about 1000feet [300m] lower down the pass."

Hybridizers have graciously spared using it, perhaps because they have felt that Nuttall was more honest than Hooker, when he named it *R. sparsiflorum*! But then, perhaps they haven't discovered the Dochu La form?

R. camelliiflorum Hooker f. "with flowers like a camellia"

a

b

eiv

oi

e

ciii

cii

gii

hiii

f

li

kiii

SUBSECTION **CAMPYLOGYNA** (Hutch.) Sleumer
(Series *Campylogynum*)

ROBERT J. MITCHELL

*R*hododendron campylogynum with its thick, bent style, is one of the most popular of the dwarf species and is widely grown. It suits the smallest of gardens: it is easy to grow, produces flowers in profusion and is very amenable in cultivation. In the cool Scottish summer this is a plant for the sunny exposures, conditions very like those found in the wild. It grows best in the lower rainfall areas of Britain.

This plant, first discovered by Abbé Delavey on the Tali (Cangshan) Range in western Yunnan in 1883, has a wide distribution. Its present known localities extend from the main centre of collections on the west bank of the Yangtse River in Lichiang and North-West Yunnan south to the Cangshan Range and west to the Mekong/Salween Divide and into North Burma. In the north it grows in East, Central and South-East Xizang (Tibet) and over the border into Arunachal Pradesh (Assam) in India. It is often the dominant species on the highest of mountains at an altitude of 3355–4700m (11,000–15,500ft), although Frank Kingdon-Ward actually found one community growing as low as 2400m (8000ft) in a cold valley in the shadow of Imaw Bum in eastern Burma. Generally it is a plant of the high ridges and open ground or one that grows among scrubby vegetation.

Due to its wide distribution and great variety of habitats, it is therefore not surprising that it is a very variable plant in cultivation. It is recorded as mat-forming to 1.8m (6ft) tall in the wild, with flower colours ranging from white, pink, red, and magenta to purple-black. The leaves have varying degrees of scaling as well as colour on their undersurfaces. Only one species with no subspecies or varieties is recognized by the new rhododendron classification, yet 20 different names are published in horticulture and are offered by the trade in Britain.

My favourite plant was grown by Betty and George Sherriff at Ascreavie as *R. campylogynum myrtilloides*, with plum-coloured, single flowers densely covering the plant, which was close on 60cm (2ft) across and less than 22cm (9in) tall.

It was as a compact, almost mat-forming plant, no more than 10cm (4in) tall, growing on open ground among rocks, that I first encountered *R. campylogynum* in the wild, at 3800m (12,500ft), just below the highest peak of the Cangshan Range. This plant – SBEC 519 – with its claret-coloured flowers and dark green, shiny foliage, is proving to be a good garden plant.

Since its introduction, *R. campylogynum* and its various forms have attracted attention and have rightly been recognized as good garden plants. A pink-flowered plant grown by Lionel de Rothschild from the Kingdon-Ward collection (3172) of *myrtilloides* from the lowest known altitude of 2400m (8000ft) was given an Award of Merit (A.M.) in 1925. Its enhanced merit as a good garden plant was acknowledged by the award of a First Class Certificate (F.C.C.) in 1943 to Edmund de Rothschild.

Collingwood Ingram was granted three Awards of Merit in 1966, 1973 and 1973 for his cultivars of *R. campylogynum*: 'Thimble' with salmon-pink flowers, 'Baby Mouse' and the white-flowered 'Leucanthum', while Lord Aberconway received the A.M. in 1975 for his 'Beryl Taylor', a selection from the collection of Ludlow, Sherriff and Taylor (4738).

The type plant of *R. cremastum* (also called *R. campylogynum* Cremastun Group) was collected by Forrest in 1917 as a 60cm (2ft) tall plant at an altitude of 3355m (11,000ft) in an open situation among rocks and dwarf scrub. In gardens this plant is erect in stature, but it is the pale green colour of both leaf surfaces and the large, bright red flowers which distinguish it.

The type plant collected by Farrer (1670) features in the *Botanical Magazine* in 1935 as variety *charopoeum* (also called Charopoeum Group). It was collected in North Burma and is abundant in North-West Yunnan, where it grows on alpine meadows. In horticulture this is quite a distinct plant with its larger flowers (to 2.5cm/1in) long) and its tiny calyx lobes.

Forrest collected var. *celsum* (also called Celsum Group) on the Cangshan Range where at 3300m (11,000ft) it grew to 1.8m (6ft) tall in open areas near to bamboo thickets. The height factor and its bushy habit identify it from the others.

Other varieties on offer give an extended range of flower colour and plant heights, from 'Crushed Strawberry' to 'Hillier Pink', 'New Pink', 'Claret', 'Patricia' (a recent introduction from the United States growing to 60cm/2ft with plum-red-coloured flowers) or 'Esther Berry' (dwarf with plum pink flowers).

Whatever your needs, the range of plants encompassed in *R. campylogynum* will provide interest and the joy is that they all flower in their very early years and require very little attention.

R. campylogynum Franch.

wān zhù dùjuān "bent-style rhododendron"

a

gv

b

o   e   d   hiii   gii   ci

k   kiii

li

kix

kxi

SUBSECTION **CAROLINIANA** (Hutch.) Sleumer

(Series *Carolinianum*)

SONJA NELSON

Defying the norm in the world of rhododendrons is *Rhododendron minus*, a native of the south-eastern United States, which thrives in temperature extremes and eschews the mild, moist climates most of the genus loves.

Standing alone in the Subsection *Caroliniana* of Section *Rhododendron*, the species *R. minus* is represented by two varieties, *R. minus* var. *minus* and *R. minus* var. *chapmanii*. In addition, the variety *R. minus* var. *minus* embraces certain horticulturally different members in what is now called the Carolinianum Group. These rhododendrons were formerly given species status under the name *R. carolinianum*. The taxonomic puzzles with which these plants have challenged botanists reflect the wide variation within the species. Appreciation of this enigmatic rhododendron requires recognition of its valuable characteristics, which are unlike those of other rhododendrons, as well as affection for the fine qualities it shares with the rest of this horticulturally and aesthetically outstanding genus.

The homeland of *R. minus*, the south-eastern United States, is an area that is uniquely rich in its flora. The diversity of plants is thought to be a result of the southern retreat of plants to escape glaciers during the Pleistocene era. It is therefore possible to find plants of both northern and southern origin growing next to each other. It was this botanically diverse region that so excited the early plant explorers. The introduction of *R. minus* is ascribed to the Frenchman André Michaux, who came upon it in North Carolina in the Appalachian Mountains as early as 1796, while collecting for his government in the wake of the American Revolution. It was described in his *Flora Boreali Americana*, published in 1803. Following in the footsteps of Michaux through the Appalachian Mountains, Asa Gray in the 1840s collected what was later designated by Alfred Rehder in the *Rhodora* of 1912 as the type specimen of *R. carolinianum*, later still changed to *R. minus* var. *minus* Carolinianum Group.

Rhododendron minus grows in the wild in North Carolina, South Carolina, Tennessee, Georgia, Alabama and Florida, from the Appalachian Mountains of the Carolinas to the lowland tributaries of the Chattahoochee River in Georgia and Alabama. Defying all odds, *R. minus* var. *chapmanii* grows in the sand, in the hot coastal plain of northern Florida.

In keeping with its "individualism", *R. minus* is the only lepidote, besides *R. lapponicum*, native to the United States. Its neighbours in its native habitat include elepidote rhododendrons and deciduous azaleas.

The variation in the *R. minus* habitat is equalled only by the variation in its form. In height it can vary from 1.2–6m (4–20ft) and more, and its growth habit can vary from upright and straggly to the shorter and compact members of the Carolinianum Group. The leathery and scaly, evergreen leaves too can vary from dark to paler green in colour and from 2–12cm (¾–4¾in) in length, and the flower colour of the small, dainty trusses ranges all the way from white to pink to pale rosy-purple. Recently, superior pink forms have been found in Alabama and Georgia and are being propagated for the nursery trade.

Although loved by those who know *R. minus* in the wild, this species truly comes into its own as a parent. Its cold and heat tolerance – it is rated as hardy to -32°C (-25°F) – has been used in crosses for years by many of the world's great hybridizers. It is parent to 'Dora Amateis', was awarded the First Class certificate (F.C.C.) in 1981 and is still a favourite among gardeners. Along with *R. dauricum*, it is also a parent to the indestructible PJM Group of cultivars which thrive in the gardens of Maine and Nova Scotia. Newly registered hybrids with *R. minus* as parent are frequently added to the *International Rhododendron Register* (recent 1994 additions include: 'Joseph Dunn', 'Carolina Gold' and 'Carolina Nude').

Any success in the garden cultivation of *R. minus* depends, not surprisingly, upon climate and soil. The British Isles and the Pacific North-West climates, so hospitable to a great many rhododendrons, are less than optimum for this species, whereas in south-eastern and north-eastern states in the United States, which experience extreme high and low temperatures, *R. minus*, in one or another of its variations, is considered a "good doer". The main requirement for soil is very good drainage, and it is best for the plant to be sited over stones or gravel.

The most exciting displays of the maverick *R. minus* I have seen are the cliffhangers along the Blue Ridge Parkway in North Carolina. However, the Rhododendron Species Foundation in Federal Way, Washington, also has several specimens of distinction. The dainty trusses against a contrasting foil of heavy-textured leaves make for a most elegant statement of spring.

R. minus var. *minus* Michx. "smaller"

a

k

bi

ci

kv

mi

b

ci

gii

hiii

d

g

Subsection CINNABARINA (Hutch.) Sleumer

(Series *Cinnabarinum*)

KENNETH J.W. LOWES

"Of hardy lepidote rhododendrons, they are almost the most beautiful," wrote Euan H. M. Cox and his son Peter A. Cox in their book *Modern Rhododendrons*. The remark refers to forms of *Rhododendron cinnabarinum*, depicted opposite. I am sure that many other enthusiasts for the genus agree – certainly I do. That beauty lies in the eye of the beholder is a proverb so deep-seated in the language that by now it has become a cliché, sometimes containing the suggestion that beauty cannot or should not be defined or explained. The beauty of *R. cinnabarinum*, however, need not suffer any such restriction, although different individuals may explore a wide range of vocabulary to express themselves satisfactorily.

This extraordinary species possesses a range of outstanding qualities, easy to list but difficult to equal. Perhaps I am most attracted by the particularly graceful inflorescence, composed of a cluster of slim bells shaped and suspended rather like those of the equally beautifully lapageria. The colour range is itself extraordinary, but the limitations of language confine me to cinnabar, crimson-purple, petunia, plum-red, purple, copper, orange, yellow and innumerable combinations of these.

In the growth of the shrub there is an overall elegance of port which agrees notably with that of the inflorescence and even the leaves. The corollas vary in shape and dimensions from one individual plant to another, always having the general appearance of bells, almost tubular and often flaring at the mouth to a greater or lesser extent. The slender growth habit is noticeable in younger plants, but many more mature specimens have thickened into dense bushes, impenetrable, yet still revealing the elegance always hinted at in flower and leaf. The relatively small, evergreen leaves, of many variations in length and width, have a healthy-looking, glossy, dark green surface. Beneath, there are variations in scaliness and in the grey to purplish-grey tint the scales produce. Nearly all forms of *R. cinnabarinum*, of whatever provenance, are free-flowering. Many are quite prodigal, and this profusion is repeated year after year if the plant is well suited in site and soil.

I have mentioned variations in flower and leaf. This is to be expected in view of the history and sources of our own plants. The species was discovered in Sikkim by Sir Joseph Hooker in the spring of 1849, and introduced to Britain without delay. Since then it has been found and reintroduced by several collectors over a wide territory including Nepal, Bengal, Bhutan and nearby China, and at varying heights. Variations in distribution area, locality and appearance are of interest to both taxonomist and gardener when a plant is important.

Most forms of *R. cinnabarinum* are reasonably hardy in recognized rhododendron-growing situations. The desirable formula of woodland or near-woodland conditions, with dappled shade and freedom or protection from heavy winds and severe frost, should be provided as nearly as possible.

The species about which I am enthusing has a few relatives, most of them desirable garden plants for some of the same reasons. The Subsection *Cinnabarina* consists of only two species, the other one being *R. keysii*. The species *R. cinnabarinum* is itself split into three subspecies, the other two being *xanthocodon* and *tamaense*. The last-named was only discovered in 1953 by Frank Kingdon-Ward. It is more or less deciduous, with flowers ranging between lilac and purple. Subspecies *xanthocodon* exists in a number of variations, with generally shorter and wider bells than subspecies *cinnabarinum*. Colours range from primrose and apricot to lavender-pink and purple.

In the wild, the many forms mix and hybridize – so it is not surprising that gardeners have exploited this tendency. The most useful discovery, many years ago, that the various *Cinnabarina* mated very readily with members of the two large Subsections *Maddenia* and *Triflora*, has borne prolific fruit. There are now in cultivation many hybrids of exceptional beauty and usefulness, combining in ever-increasing diversity the best characteristics of these three subsections. Yet – and perhaps inevitably – I am so deeply under the spell of *R. cinnabarinum* ssp. *cinnabarinum* that it would displace any of them.

R. cinnabarinum Hooker f.

zhū shä dùjuän "cinnabar rhododendron"

a

b

ci

ki

h*iii*

d

e*i*

g*ii*

g*v*

g*i*

h*iv*

k

l

SUBSECTION **EDGEWORTHIA** (Hutch.) Sleumer

(Series *Edgeworthii*)

ROY LANCASTER

*T*he bamboo thicket was unending and the track we followed constantly changing as animal trails crossed or merged. There were eight of us, three British and five Chinese, members of an expedition to the Cangshan (Dali Mountains) of western Yunnan in May 1981, and we had made the decision against our guide's advice to persist with a track which increasingly appeared to be going nowhere. In the middle of the day it would not have mattered, but it was early evening and we were a long way from camp with the sun on the wane. We were all privately thinking it was time to turn back when we detected a delicious fragrance sneaking through the bamboo canes.

Following our noses we pressed on and in a short while broke through to where a series of rock slabs angled steeply into the mouth of a ravine. The ravine was deep but we could distinctly hear the roar of the torrent in its belly. The fissures between the rock slabs were filled with grass and small herbs in which were embedded substantial clumps of two terrestrial orchids, *Pleione bulbocodioides* and *P. forrestii*, with rich purple and golden-yellow blooms respectively. Beautiful though these were, they were not the source of the fragrance. Then we saw the bold, trumpet-shaped flowers of a small, puckered-leaved rhododendron.

It was my first meeting with *Rhododendron edgeworthii* in the wild and it is a moment I shall never forget. First described by Joseph Hooker in the Sikkim Himalayas in 1849, it is named after Michael Packenham Edgeworth, an Irishman employed by the Bengal Civil Service, who made plant collections in India (including the Himalayas) and Ceylon. It enjoys a wide distribution in the wild, from India (Sikkim, West Bengal, Arunachal Pradesh) through Bhutan and East Burma into China (North, North-West and Central Yunnan, South Xizang [Tibet]). My most recent experience of *R. edgeworthii* in the wild was in the mountains of central Bhutan where it was not uncommon as an epiphyte perched high in the moss-mantled crowns of evergreen oak and magnolia as well as on rocks and cliffs. Not surprisingly, it is very variable in size and habit, so much so that perceived variations have twice been given species rank – the most important of these was *R. bullatum*, described by the French botanist Adrienne Franchet from a plant collected by Abbé Delavay, in the Cangshan in 1886.

The epithet *bullatum* well describes the characteristic bullate (blistered or puckered) upper surface of the leaf. Equally significant is the densely scaly lower surface which is further covered by a dense woolly pelt of hair. This pelt also clads the petioles and branchlets, extending even to the rachis and pedicels of the inflorescence. The colour of the hairs varies from rust to a pale fawn but is typically a rich reddish-brown when mature. When newly emerged, the leaves are a soft grey-green with a light buff undersurface and a white, downy upper surface.

The large flowers in mid- and late spring are borne in trusses of usually 2–3 each, with a flared mouth, deep rose in bud opening white with delicate pink or rose-pink hues and a delicious fragrance. The calyx is quite large and often red-tinged.

In its native forests *R. edgeworthii* occurs at an altitude of 2100–3300m (6900–10,800ft) on rocks or cliffs and just as commonly as an epiphyte on trees or stumps, where it revels in the moss cover and the cool moist conditions provided by cloud and mist. There it can reach 2.5m (8ft), with long bare stems.

Given its rather exacting habit, its successful cultivation is not child's play. A combination of perfect drainage and a growing medium rich in organic matter is only part of the answer. The plant's ability to survive cold winters and spring frosts is quite another problem and, although traditionally the milder, moister western seaboard of the British Isles has provided the best sites for this species, there are hardier (Chinese) forms worth trying in more sheltered gardens inland, possibly against a wall. In colder areas, it makes a useful conservatory or cool greenhouse subject, in a pot or raised bed. Indeed, it is in such places that many of the hybrids of *R. edgeworthii* enjoy popularity, although some of these too can be grown outside in sheltered gardens. 'Lady Alice Fitzwilliam' and 'Fragrantissimum', the two oldest, best known and most reliable, are especially suitable for this – both received the R.H.S.'s coveted First Class Certificate (F.C.C.) More recent hybrids include 'Cowbell', 'Actress', 'Bert's Own' and 'Princess Alice'. The last named, together with 'Fragrantissimum', 'Lady Alice Fitzwilliam' and the incomparable *R. edgeworthii* itself, are included in the R.H.S.'s recently published list of Award of Garden Merit Plants – the ultimate accolade.

Subsections *Fragariiflora* and *Genestierana*, see pages 104–5.

R. edgeworthii Hooker f. (after M. P. Edgeworth, 1812–1881, Bengal Civil Service)

pàopào yè dùjuān "bubble-leaf rhododendron"

a

c*i*

e*i*

g*ii*

g

d

b

k*i*

h*v*

k

SUBSECTION **GLAUCA** Sleumer
(Series *Glaucophyllum*)

MERVYN S. KESSEL

The exact meaning of *Glauca*, or indeed "glaucous", depends on your source of information: William T. Stearn's *Botanical Latin* indicates that it is a light sea-green, although most rhododendron literature describes "glaucous" as a grey or blue-grey, waxy bloom. H. H. Davidian's classic work, *The Rhododendron Species*, recognized eight species and at least two varieties within the *Glaucophyllum* subseries, while J. Cullen in the *Notes from the Royal Botanic Garden Edinburgh* attributed six species, and some subspecies and varieties to the Subsection *Glauca*.

Regardless of the status of the species, this section contains many plants eminently suitable for the small garden. Apart from their relatively small and interesting flowers, most of the species in this section are noted for their aromatic foliage, which is particularly apparent when gently rubbed. Two species of this subsection (illustrated opposite) are readily available from both the specialized nursery and many of the better garden centres.

Rhododendron glaucophyllum is a native of the Sikkim Himalayas, with its distribution extending to Nepal and Bhutan, while *R. charitopes* ssp. *tsangpoense* occurs in China and South-East Xizang (Tibet). Both species grow at an altitude of around 3000m (9800ft), although frequently considerably higher.

Do not think, however, that you have to live on a mountainside to cultivate these species, for in reality they are fairly accommodating plants. Their limited size makes them ideal for peat walls or for the larger rock garden, but even if you have only a patio area, a large tub will be sufficient to grow both these species satisfactorily – provided you keep them well watered and fed.

Rhododendron glaucophyllum will reach a height of about 1m (3½ft), although in favourable gardens 1.5m (5ft) is not unusual. This species is evenly balanced and attractive in appearance, with pinkish-purple, campanulate flowers appearing in the spring, often from mid- to late spring, depending on the season. This unfortunately means that the blooms are occasionally frosted. Peter A. Cox, in his book *The Cultivation of Rhododendrons*, considers that *R. glaucophyllum* is hardy to about -18°C (-0.4°F), although,

fortunately, in Argyll I have never experienced such Arctic conditions to confirm his statement, but I am sure he is correct! Although fairly tough, they do not like drying out and cold-wind desiccation is more likely to cause damage than frost alone.

Rhododendron charitopes ssp. *tsangpoense* (named after the Tsangpo area of Xizang) flowers somewhat later than *R. glaucophyllum* and the blooms are therefore less likely to be damaged by frost. The flowers are purplish and campanulate in shape. This species was first discovered by F. Kingdon-Ward in South-East Xizang in 1924, and has gained considerable poularity due to its relative ease of cultivation and the fact that several commercial growers are making the plant readily available through garden centres.

Both of the above two species propagate fairly easily from semi-ripe cuttings, and it is worth ensuring continuity of stock, just in the off-chance that the one-in-fifty severe winter occurs when you are least expecting it. Like many other species, both *R. glaucophyllum* and *R. charitopes* ssp. *tsangpoense* are slightly susceptible to rhododendron powdery mildew, and should the problem become acute, spraying with one of the recommended fungicides may be worth considering since they are both relatively small plants, even at maturity.

Unlike many other species, *R. glaucophyllum* and for that matter *R. charitopes* ssp. *tsangpoense* do not appear to have been used very much in hybridization. There are very few recorded crosses and even fewer hybrids available in the trade. I am sure that someone will tell me that they have bred dozens of them, but so far they do not appear to have emerged. I can only assume that there is little point in trying to improve on the perfection of the true species.

Clockwise from top left:
R. glaucophyllum Rehd. "with bluish-grey leaf";
R. luteiflorum Davidian "with yellow flowers";
R. charitopes ssp. *tsangpoense* Balf. f. et Farr.;
ya róng dùjuān "pretty-face rhododendron"

k

kvii

kvii

k

a

a

hv

b

ci

ei

ci

d

b

fiii

a

b

ci

b

ci

hv

d

ei

hv

d

gii

kvii

mi

g

ei

SUBSECTION **HELIOLEPIDA** (Hutch.) Sleumer

(Series *Heliolepis*)

RICK PETERSEN

*I*n the rugged, mountainous areas of south-central China and northern Myanmar, Burma, one finds the greatest diversity of rhododendron species. Over 70 lepidote species alone are concentrated in the region where the political boundaries of Sichuan, Yunnan, Xizang (Tibet), and northern Myanmar come together. Two species of the *Heliolepida* subsection, a small group of four closely related species within the genus *Rhododendron*, are found in this area. These two are *Rhododendron rubiginosum* and *R. heliolepis*, which are the better known of the group, with *R. bracteatum* grown less often and *R. invictum* not in cultivation.

The first recorded collection of *R. rubiginosum* by a European was in 1886 on Tsang-Chan Mountain by Abbé Jean M. Delavay, a French Catholic missionary stationed in western China. Plants of this species were described and named in Paris, France, by Adrienne Franchet in 1887. The specific epithet "rubiginosum" means "reddish brown" in Latin and refers to the dark, scale-covered leaves. It was subsequently collected by plant explorers George Forrest, Frank Kingdon-Ward, Joseph Rock and others in the early part of this century. More recently, new specimens have been brought into cultivation by Peter Cox, Warren Berg, Bill McNamara, and others.

This species is native to north-western Yunnan, south-western Sichuan, south-eastern Xizang, and northern Myanmar and is found at 2290–4270m (7500–14,000ft) elevation. The growth habit of *R. rubiginosum* is a spreading to upright shrub or small tree from 60cm to 9m (2 to 30ft). The flowers may be pink, rose, lavender-rose, mauve, rose-purple to white-flushed-pale pink with crimson, purple or brown flecks. The time of flowering is usually early or mid-spring. The undersurfaces of the leaves are covered with scales. These tiny plant structures regulate water loss from the leaf and with the naked eye they look like dots. The density of the scales is variable and many forms have scales that are thick and overlapping, giving a rusty or red-brown colouring, hence the specific epithet.

In the wild, *R. rubiginosum* grows in thickets, mountain pastures, on granite boulders, limestone cliffs and beneath spruce, pine and oak. This species has a relatively wide native range, and, when first collected, several different forms were described and named, including *R. desquamatum*. Later, taxonomists decided that these species were actually morphologically equivalent to *R. rubiginosum*, and following their discoveries they were merged under that species name.

The Royal Horticultural Society gave the Award of Merit (A.M.) to the former *R. desquamatum* in 1938 and to a named cultivar, 'Wakehurst', in 1960. The latter is now considered to be a hybrid. Other named forms of *R. rubiginosum* include 'Finch' and 'Rosey Ball'. There is a dearth of hybrids using *R. rubiginosum* as a parent, with only one found by this author and that is 'Eleanore' (*R. rubiginosum* Desquamatum Group × *R. augustinii*), which received the A.M. in 1943.

In 1948, Joseph Rock wrote a letter from Likiang, Yunnan Province, China, to George Grace, Secretary of the American Rhododendron Society, which was printed in the *ARS Quarterly Bulletin* that same year. Dr. Rock had sent rhododendron seed to the Society and in his letter he described the plants from which the seed was obtained, including a description of *R. rubiginosum*: "Tree 15–20ft [4.7–6m], leaves broad, red beneath, flowers purple, in Spruce forest and open slopes in scrub forest." Peter Cox describes an expedition to the Cangshan mountain range in Yunnan in the Spring 1982 *ARS Journal*. He relates seeing *R. rubiginosum* above the base camp situated at 2500m (8200ft). Also seen in the area were *R. yunnanense*, *R. facetum*, *R. anthosphaerum*, plus an unexpected population of *R. sinogrande*.

Rhododendron heliolepis is a variable species with two varieties recognized now: var. *heliolepis* and var. *brevistylum*. They are native to the same general area as described for *R. rubiginosum* and are found at 2400–3800m (8000–12,500ft). In the woodlands and on cliffs of central Sichuan, *R. bracteatum* is found at 2100–3500m (7000–11,500ft). *Rhododendron invictum* is the least known of the subsection and is found in central Gansu Province, which is on the north side of Sichuan Province.

Of the species in the *Heliolepida* subsection I would definitely recommend *R. rubiginosum* as the easiest and best ornamental plant. In early spring these beautiful shrubs are covered with delicate blossoms. I encourage everyone who has an opportunity to visit the Rhododendron Species Botanical Garden, here in Seattle, U.S.A., to see these and many other outstanding plants.

R. rubiginosum Balf. f. et Forr.

chá huā yè dùjuān "camellia-leaved rhododendron"

a

b

ci

d

ei

g

gii

hiii

ki

kv

k

Subsection **LAPPONICA** (Hutch.) Sleumer

(Series *Lapponicum*)

Dr. Björn Aldén

The fascinating group of dwarf rhododendrons collectively known as Subsection *Lapponica* takes its name from the type species of the group *Rhododendron lapponicum*. Carl von Linné first described this species from Swedish Lapland in 1753, but placed it in the genus *Azalea*, mainly because of its short-lived leaves.

Until the botanical explorations of South-West China at the end of the nineteenth century, *R. lapponicum* had few known close relatives. It was then found that the centre of diversity of the group (as with so many other plants) is actually the Sino-Himalayan region.

Disagreement among taxonomists makes the number of species in the subsection uncertain, although there are likely to be between 20 and 30. Except for the bristly *R. setosum* – which is probably not a 'Lapponica' anyway – and the sturdy *R. cuneatum*, the group is fairly uniform, comprising small- to medium-sized shrubs with small, entire leaves and one- to several-flowered, umbel-like inflorescences, with small and broadly funnel-shaped flowers.

Differentiating between Subsection *Lapponica* and related groups such as Subsections *Rhododendron, Heliolepida, Triflora* and Section *Pogonantha* involves more technical features. Members of the last section are recognized by their typically lacerate (not crenate or entire) leaf scales. From Subsection *Rhododendron* the 'Lapponicas' differ by having entire (not crenate) leaves with microscopic, epidermal processes. The latter character can also be used to distinguish Subsection *Lapponica* from Subsections *Triflora* and *Heliolepida*. Of the lepidote rhododendrons (subgenus *Rhododendron*), Subsection *Lapponica* has the widest distribution, due mainly to *R. lapponicum* itself, which is almost circumpolar.

One of my first encounters with this species was late one night at the end of July, in a scenic area around Abisko in North Sweden. Here, well above the arctic circle, it grows on calcareous mountain slopes and in bogs. Not more than 5cm (2in) high and hardly lifting its rosy-red flowers above a mat of *Cassiope*, it was a while before I spotted it. However, the tiny, rosy-red flowers, glowing in the midnight sun even allowed me to forget the mosquitoes.

In cultivation, *R. lapponicum* from North Scandinavia has proved exceedingly difficult. Even if it survives for more than a few years, it hardly produces any flowers. This phenomenon is typical of northern-latitude plants when transferred to lower latitudes. Fortunately, the species also occurs at more southerly latitudes and material from North-East Asia and Japan grows well

even in Central Sweden. These Asian plants are now known as *R. lapponicum* Parvifolium Group.

One of the merits of Subsection *Lapponica* lies in its variation in flower colour. In addition to the most frequent purple to blue, red or occasionally white, there are also a couple of yellow-flowered species. I have had the pleasure of studying one of these, *R. rupicola* var. *chryseum*, in North-West Yunnan where it is common in alpine habitats. Further north, in the mountains around Kanding in West Sichuan, another yellow-flowered species occurs. This was discovered by Abbé Soulié and described in 1895 by A. E. Franchet who gave it the epithet *flavidum* (meaning 'yellowish').

Rhododendron flavidum is quite tall, up to 1m (3ft) in cultivation and to twice that size in the wild. Its upright twigs bear dark green, shiny leaves. Like *R. rupicola* var. *chryseum*, it can be found in the upper forest zone as well as in open alpine habitats. Several clones of *R. flavidum* still exist in cultivation from E. H. Wilson's collections of 1905 and 1908. In Gothenburg, the species did well for at least 30 years before it succumbed to the winter of 1987.

Almost as distinctive as the yellow-flowered species is *R. hippophaeoides*, which has flowers the colour of crushed blueberries in milk. It is further characterized by leaf-shape (*hippophaeoides* means "like sea buckthorn") and the pale-scaled undersides of the leaves. This species is one of the best known members of the group in cultivation, particularly in the British Isles and western North America. The best forms, with lavender-blue flowers, are simply fantastic. In the wild, *R. hippophaeoides* is especially common around the Zhongdian plateau in North-West Yunnan, where it grows in wet meadows or along rills on calcareous ground and attains a height of at least 1.5m (5ft). White forms occur very occasionally but despite several efforts (my own included), they have yet to be introduced to cultivation. The most noteworthy hybrid is a *Triflora*-cross called 'Mother Greer', a compact, early-flowering shrub with lilac-blue flowers.

Subsections *Ledum* and *Lepidota*, see page 105–6.

Clockwise from top:
R. lapponicum (L.) Wahl. (from Lapland);
R. flavidum Franch. dàn huáng dùjuān "light yellow rhododendron";
R. hippophaeoides huī bèi dùjuān "grey-backed rhododendron"

a

a

a

ci

b

hiii

ci

d

gii

k

k

b

k

k

d

e

hiii

hv

hv

Subsection **MADDENIA** (Hutch.) Sleumer

(Series *Maddenii*)

G. Alan Hardy

*M*ajor Madden, traveller, botanist and studier of the tropical vegetation on the western Himalayas, must have been very pleased and proud in 1849 when Sir Joseph Hooker named *Rhododendron maddenii* after him. Madden must rate in history as having had his name appended to one of the most beautiful and yet most complex series of rhododendrons. It has probably the widest distribution of any series, spreading from Nepal to Burma and Yunnan, Thailand, Laos and Vietnam, but it is in China where most are found. To show further its complexity, there have already been four revisions. The first two by Hutchinson recognized some 46 species, later reduced to 21 by Sleumer. The last, by James Cullen in 1980, recognized 36 species divided into four groups which more or less correspond to Hutchinson's three.

If we now briefly look at the four groups, we find that Group 1 has only *R. maddenii* itself and *R. crassum*. These two are very closely integrating and are the hardiest members of the series. Group 2, or the "Dalhousiae Alliance", without doubt has the most exciting and distinguished members, including our subject, *R. lindleyi*. No one can fail to be somewhat overawed by *R. nuttallii*, which was, for me, an early love. *Rhododendron megacalyx* with its wonderful nutmeg scent holds the fort alone in Group 3. Group 4 is the largest and most variable. All previous alliances have all white rhododendrons, but here we see the introduction of pink and yellow and also the smallest-flowered member in *R. valentinianum*. Again in this group are one or two of the hardier members – *R. ciliatum* and *R. johnstoneanum* being those that will survive outside. Scent in this group is also notable, *R. veitchianum* in particular possessing a most powerful scent.

It is disappointing therefore that these rhododendrons cannot be grown much out of doors in Britain. They are, however, very amenable plants in a cool greenhouse or conservatory which should remain just above freezing if the plants are to flower well. When grown in containers, vine weevils can be a nuisance.

Rhododendron lindleyi has had a somewhat chequered career from the start. For a long time it was confused with the Dalhousiae, even by Hooker, its discoverer, in 1848, but it finally fell to Thomas Moore in 1866 to describe the species from a plant growing in Standish's nursery at Ascot. That particular one was reputedly collected by J. J. Booth, who was working in Bhutan from 1849 to 1860. An interesting account of the mix-up is recorded in *RBG Notes* Vol. 12 of 1991. Little was heard of fresh introductions until 1936 after Ludlow and Sherriff reintroduced *R. lindleyi*. *R. lindleyi* appears under the number 1205 in the R.H.S. handbook, but of a very distinct form. Until then, *R. lindleyi* was white, but here it was strongly flushed with red purple. It also had a different bud colour to the usual green. David Davidian in his excellent book called this *R. grothausii*, but in Cullen's classification it is still *R. lindleyi*. This is the plant known as *R. lindleyi* 'Geordie Sherriff'. Fortunately, both Tom Spring-Smythe and Peter Cox later reintroduced good Maddenias, and probably Peter Cox's *R. lindleyi* is the best form of any.

Rhododendron lindleyi has a wide distribution and, in the wild, is usually an epiphytic shrub up to 5m (16ft) high but slightly less in cultivation. The branchlets are moderately or densely scaly with olive-green, oblong lanceolate leaves 7.5–13cm (3–5in) long and some 5cm (2in) wide, which are reticulate above with the underside a glaucous green and sparingly scaled with scales of unequal size. The flower buds are green and covered with rusty scales. Flowers are white or white-tinged-pink on the margin of the corolla lobes, generally with a yellow blotch and fragrant.

Rhododendron lindleyi has received several awards: in 1935 an Award of Merit (A.M.) to Lionel de Rothschild, Exbury, with flowers flushed rose-magenta; in 1937 a First Class Certificate to Admiral A. Walker-Heneage-Vivian of Clyne Castle, Swansea, with flowers tinged pink at the end of the corolla lobes; in 1965 to Geoffrey Gorer, Haywards Heath for a clone 'Dame Edith Sitwell' with flowers white-tinged-pale-pink; and in 1969 an A.M. to A.C. and J. F. A. Gibson of Glenarn, Scotland, for a clone 'Geordie Sherriff' with flowers flushed externally with red-purple.

I hope I have done justice to *R. lindleyi*, but perhaps a comment written to me in 1988 by Mrs. Jean Rasmussen, the late Capt. Kingdon-Ward's wife after the London rhododendron show, says it all: "Why did the Lindleyi have no scent? – In the forest it can betray itself by its scent alone even without seeing those glorious white trumpet flowers." What better epitaph!

R. lindleyi T. Moore (after Dr. J. Lindley, 1799–1865, a former Botanist and Secretary to R.H.S.)
dà huā dùjuān "large-flowered rhododendron"

a

b

ci

f

d

hiii

ei

j

gv   gii

k

SUBSECTION **MICRANTHA** (Hutch.) Sleumer

(Series *Micranthum*)

HAROLD GREER

"Unusual in the rhododendron world" is an understatement when referring to *Rhododendron micranthum*. This lepidote in Section *Rhododendron* is the *only* plant in the Subsection *Micrantha*. It does not hybridize readily with other rhododendrons and, as far as I can ascertain, there are no known hybrids. It appears more closely related to the genus *Ledum* than *Rhododendron*, although in the eyes of a few botanists, plants in the genus *Ledum* are rhododendrons.

As an aside, I might mention here that the major difference between the genus *Ledum* and the genus *Rhododendron* is the fact that *Ledum* opens its seedpod backwards. In all rhododendrons, the seed opens at the end of the seedpod where the style protrudes. In *Ledum* the seedpod opens on the back end where it is connected to the pedicel. Furthermore, there are one and maybe two supposed crosses between *Ledum* and *Rhododendron*. The first is the plant 'Arctic Tern', which was originally thought to be the species *R. trichostomum* var. *ledoides*, but has since been declared to be a hybrid of *R. trichostomum* var. *ledoides* and *Ledum*. The second possible hybrid is the plant × *ledodendron* 'Brilliant', which was said to be a hybrid of *Ledum* and *R.* 'Elizabeth', though, some experts consider it a hybrid of *R.* 'Elizabeth' and an evergreen azalea. Just how does all this relate to the topic of *R. micranthum*? Well, *R. micranthum* is very much like a *Ledum*. It has trusses with a large number of flowers and a long rachis, which most other lepidotes do not have. Wouldn't it be great to have other lepidotes that have larger flowers, with 20 or 30 flowers to the truss?

Rhododendron micranthum was first described by Turczaninov in 1837, from a specimen collected on the mountains north of Peking, and was illustrated in the *Botanical Magazine* t. 8198 in 1908. Its name in Greek means "small flower", and in Chinese it is known as "shining mountain white" or "white mirror". It is widely distributed and grows wild from North Korea through central and northern China and Manchuria. It grows in the wild at elevations between 1500 and 2500m (5000 and 8000ft) appearing in thickets, on cliffs, in dry gorges and on ridges. This species was first introduced by Wilson from western Hupeh in 1901, and was reintroduced by him again in 1907. It is said to be intensely poisonous, especially the young leaves, which can kill livestock unlucky enough to eat the plant.

Rhododendron micranthum, on casual observation, looks more like a spiraea than a rhododendron. It flowers in early summer after most rhododendrons are over, and can easily be mistaken for a plant other than a rhododendron. It generally grows to 1.2–2.5m (4–8ft) tall by about the same width.

I have known this plant since the early 1960s when I obtained a plant from the late James Caperci of Seattle, Washington. It is an interesting plant, with large flower buds that by late autumn look as if they cannot wait until spring to burst forth. But they do wait, not showing colour until early summer or even midsummer. This species has willowy limbs which never become strong branches, and form instead a loose mound of distinctive foliage with buttery-white flowers. There is a fair amount of variation between cultivars of *R. micranthum*, with some having longer leaves, more compact growth, or more mid-winter flower buds. This is not a common rhododendron in cultivation, although it is not difficult to grow in the parts of the world that are considered favourable to growing rhododendrons. It is hardy to at least -20°C (-4°F). This is an unusual plant, worthy of a position in your garden.

Subsection *Monantha*, see page 109.

Subsection *Monantha*, see page 109.

Editor's note Genus *Ledum* is now included with *Rhododendron* (Kron & Judd, 1990). Ref. R.B.G. Edinburgh. Harri Harmaja. Taxonomic notes on Rhododendron subsection *Ledum* (*Ledum*, *Ericeaceae*) with a key to its species Ann. Bot. Fennici 28:171-73, 1991. See also page 105.

R. micranthum Turcz. "small-flowered"
zhào shān bái "shining mountain white"
bái jingzi "white mirror"

a

b

jii

d

ei

r

oi

kx

kvii

ci

kiii

hii

k

gii

SUBSECTION **MOUPINENSIA** (Hutch.) Sleumer

(Series *Moupinense*)

Dr. Noel Sullivan

*W*hat a pleasant surprise to have a preview of spring in your garden on a winter's day! This will happen if you cultivate *Rhododendron moupinense*, a species native to western Sichuan, China, and named after the district of Moupin.

Few winter-flowering rhododendrons are white, especially among the lepidotes, and *R. moupinense*, with its basically white flowers, can make a dramatic statement, for it is so floriferous that the foliage can be completely hidden by the wide-open florets.

Rhododendron moupinense was first described by Franchet in 1886 and introduced to cultivation by Wilson in 1909. This species may be identified with leaves 2.5–5cm (1–2in) long and half as wide, oval in shape, deep green and very shiny on the upper surface, paler beneath, the margins recurved and the leaves extremely thick and stiff for their size. The inflorescence is terminal and the flowers solitary or up to three. The corolla is widely funnel-shaped with five lobes and the margins are frilled. The colour is white, white-flushed-pink or white in the tube with the remainder rose-pink, but deeper in colour on the upper three lobes. There are twin rays of spots on the side margins of the upper lobe, pale in the white form and bright crimson in the pink form.

There are three growth forms in cultivation: the most usual is an upright shrub of under 1.8m (6ft) with a spreading head and beautiful, dark brown, peeling bark and with white-flushed-pink flowers. The form with white flowers is a very compact and rounded shrub of up to 90cm (3ft) tall, and the other a lax and spreading shrub with smaller typical foliage and deep rose-pink flowers.

Despite coming from a relatively low altitude, the plant is perfectly hardy in cultivation and the buds and opened flowers are tough enough to resist a few degrees of frost, but for flowering in winter it needs to be placed away from a frost pocket, so air drainage is essential. In the wild, this species grows in woodland, most commonly as an epiphyte on oaks but also on rocks and cliffs, so its roots must also have perfect drainage. Thus it is very suitable for growing on old stumps and fallen logs, but it also does well in humus-enriched soils. Its epiphytic nature also offers high drought resistance. In colder areas, it makes an excellent pot plant which may be moved indoors during the flowering period. Thus *R. moupinense* is a unique member of this surprising genus.

In Subsection *Moupinensia* there are two other rhododendrons that come from the same general location in China: *R. dendrocharis*, which literally means "tree adorning", and *R. petrocharis*, which describes its habit of growing on rocks. Both are similar to *R. moupinense* but smaller. *R. petrocharis* has white flowers, and *R. dendrocharis* has red flowers. Neither species is in cultivation nor have they ever been introduced.

Not surprisingly *R. moupinense* has been used extensively in the production of early-flowering hybrids, and up to 1960 some 17 had been registered by well-known English hybridizers. They were so successful that most are still available in the trade, such as *R.* 'Bulbul', *R.* 'Chrypinense', *R.* 'Olive', *R.* 'Bo Peep', *R.* 'Bric-a-Brac', *R.* 'Tessa', *R.* 'Seta', *R.* 'Golden Oriole', *R.* 'Valpinense', and *R.* 'Cilpinense', which grows as a compact mound smothered with shell-pink flowers. There is an Australian-raised F_4 of this, very dwarf with light red flowers.

In the next period, from 1960 to 1980, production shifted to the west coast of North America and another seven hybrids were registered. Possibly the most popular was Bob Scott's *R.* 'Pink Snowflakes' (1963); the other parent was *R. racemosum*. *R.* 'Pink Snowflakes' is a semi-dwarf gem, superior in foliage and habit throughout the year, with multiple mahogany-red buds against glossy, deep-green foliage and flowers in late winter with a pink and white display. Bob Scott has continued to use *R. moupinense* in complex crosses with the intention of producing better pinks and reds, such as *R.* 'Scott's Valentine', *R.* 'Beverley Court' and *R.* 'Lake Merrit', with good plant habit and perfume.

In an even milder climate in north-western Tasmania similar goals have been independently pursued and the results, prior to registration, include larger flowers in trusses of up to ten and deep pinks and light reds, some with contrasting blotches.

I can wholly recommend that *R. moupinense* and its hybrids be grown as a group to brighten a garden in the early days of spring and late winter.

My interest in rhododendrons is better described as an obsession, with an equal infatuation for species and for hybrids of distinction. Formerly a private gardener I now enjoy my retirement as Curator of the Emu Valley Rhododendron Garden in Burnie, Tasmania.

R. moupinense Franch. (from Moupin, western China)
bǎo xǐng dùjuān "Baoxing rhododendron"

a

b*i*

h*v*

h

g

d

e*i*

o*i*

c*i*

k

SUBSECTION **RHODODENDRON**

(Series *Ferrugineum*)

RONALD MCBEATH

For me it is a great thrill, and I am sure it is the same for most other gardeners, to find for the first time growing in the wild a plant with which I am very familiar and which I enjoy growing in my garden. I suspect that like me, the first wild rhododendron that most Europeans will find is either *Rhododendron ferrugineum* or *R. hirsutum*. I can clearly remember the day, about 30 years ago, when climbing up a mountainside in Switzerland in an open grassy area strewn with boulders and amongst stunted, bonsai-like spruce trees, I came across my first wild rhododendron, *R. ferrugineum*! What excitement. Although they were straggly, stunted specimens that would have been consigned to the bonfire if they grew like that in the garden, it was a wonderful feeling to find my first rhododendron.

Rhododendron ferrugineum, the 'Alpenrose', is in fact the type species for the whole genus *Rhododendron* and is in Subsection *Rhododendron*. It was described by Linnaeus in 1753, although it was reported to be in cultivation in 1739. Subsection *Rhododendron* contains only two other species: *R. hirsutum* and *R. myrtifolium* (*kotschyi*).

Rhododendron ferrugineum is quite common across much of the European Alps, as far east as the western parts of former Yugoslavia, and is common in the Pyrénées. It is normally found on soils derived from acidic rocks and its altitudinal range is between 900 and 2150m (3000 and 7000ft). In favoured locations it can dominate the vegetation on a hillside. In the wild it will reach 1.2m (4ft) tall, sometimes making an open straggly bush, but in gardens it can be quite slow growing, neat and compact. The leaves are dark green and shiny on the upper surface, rusty brown (hence the epithet *ferrugineum*) on the lower surface, with a dense layer of tiny scales. The flowers can vary from deep rose through to crimson-purple, or white in var. *album*. The flowering period is not until early to midsummer, which makes this species a good addition for extending the flowering season in the rock or peat garden. In Edinburgh we find that it grows perfectly happily in our well-drained, light, sandy, acid soil in the rock garden and, to thrive, it does not require the humus-rich conditions found in the peat garden. It grows in full sun and exposure, remains compact and flowers freely most years. Propagation is reasonably easy from seed and young plants will grow quite fast in their early years.

Rhododendron hirsutum is easily distinguished from *R. ferrugineum*, as the former does not have the dense layer of brown scales on the undersurface of the leaf, and the leaf margins, petioles and branchlets are conspicuously bristly (hence the epithet *hirsutum*). This species is also easy to grow in the garden, producing rose-pink to scarlet flowers in early and midsummer. It is reputed to have been in cultivation since 1656 and was described by Linnaeus in 1753. It is mostly found in slightly calcareous soils and can even be discovered growing in cracks in limestone rocks – its natural distribution is the central European Alps, Dolomites and former Yugoslavia. Where the habitat is suitable for both *R. ferrugineum* and *R. hirsutum* to grow close together, hybrids can occur and go under the name × *intermedium*.

The third species in the subsection is *R. myrtifolium*. It is to be found in South-East Europe and resembles a small form of *R. ferrugineum*. It is less frequently seen in gardens and does not appear to grow so well or flower so freely, although it often has a scattering of flowers throughout the summer and autumn.

There are few cultivars in this subsection. White forms can be found in all three species – 'Coccineum' and 'Atrococcineum' are selections of *R. ferrugineum* with scarlet flowers, and there is a double form of *R. hirsutum*.

R. ferrugineum Linn. "rust-coloured"

gi

g

a

b

ci

d

hiii

cii

kxi

ki

kvii

ei

k

gvii

gii

SUBSECTION **RHODORASTRA** (Maxim.) Cullen

(Series *Dauricum*)

DAVID LEACH

*F*ew are the harbingers of spring that splash the landscape so lavishly with colour as do the two species with small, scaly leaves in the *Rhodorastra* subsection, *R. mucronulatum* and *R. dauricum*, both beloved for their brave heraldry in a bleak season. In a mild climate, both species may produce their 5cm (2in) mauve-pink flowers as early as mid-winter; in the north-eastern United States they flower from early to mid-spring.

Rhododendron mucronulatum was first described in 1837, introduced by Dr. Bretschneider in 1882 from China, and re-introduced by Kew Gardens in 1907 from Japan. It is a broadly upright shrub, 1–4m (3½–12ft) high, with large, deciduous leaves about 3.5–7.5cm (1½–3in) long, bright green, and with a laxly scaly underside, the leaf apex acute or acuminate. The branchlets are long and slender. Its open, airy look is charming when it is smothered with flowers, especially those with the colour purity of the popular clone, 'Cornell Pink'. It combines in a beautiful duet with yellow daffodils in its season. 'Mahogany' has dark, reddish-purple flowers; a white form is seen occasionally in Japan, but rarely in the colder parts of the United States, where most growers have seen quite enough of white in the preceding months. The Award of Merit (A.M.) has been awarded three times: in 1924 to a form shown by Kew; in 1935 to 'Roseum', and in 1965 to 'Cornell Pink'. 'Winter Brightness' received a First Class Certificate (F.C.C.) in 1957, and 'Winter Sunset' for its near-magenta flowers, but neither seems to be commercially available.

Rhodendron mucronulatum is found in so many variations because of its immensely wide distribution in nature. It has been collected in China, Manchuria, Korea, Siberia and Japan and its offshore islands, which account for its hardiness. The most notable hybrids are Joseph Gable's *R.* 'Conewago' (with *R. carolinianum*), his *R.* 'Pioneer' and *R.* 'Malta', derived from the latter by myself. Some high-elevation forms in Korea are uncomfortably close in their traits to the other species in this subsection, *R. dauricum*.

Rhododendron mucronulatum is exceptionally adaptable. It is easily grown with standard rhododendron culture in climates both cold and mild, and is reported to withstand the heat and aridity in southern California and in the mid-Atlantic states.

Much that has been written here about *R. mucronulatum* applies as well to its close relative, *R. dauricum*. *R. dauricum* was described by Linnaeus in 1753, and first introduced in 1780. It grows to 1.8m (6ft) high. It is among the earliest of all woody plants to bloom, often simultaneously with witch hazel, which makes a handsome pairing. The flowers of *R. dauricum* are about 5cm (2in) across, borne in abundance, in shades of mauve, purplish-rose or white.

The conspicuous and very important difference between the two species is the persistent foliage of *R. dauricum*, evergreen the year around, and therefore preferable in most garden situations. The thick, usually oval to oblong-elliptic leaves, are smaller, the leaf apex rounded or obtuse, the lower surface densely scaled with overlapping scales. The leaves are pleasantly aromatic, and their shape affords the easiest distinction to separate this species from *R. mucronulatum*, with its thinner, elliptic to lanceolate foliage.

My 40-year-old plants, grown first in the frigid mountains of western Pennsylvania and then in the flatlands of northern Ohio, are now 2.5m (8ft) tall, with reliable masses of mauve blossoms year after year. The flowers tolerate several degrees of frost, but very rarely are they destroyed by severe out-of-season freezes.

There is a fine Baldsiefen introduction, 'Arctic Pearl', which has white flowers and a superior growth habit; it blooms several days later than the typical mauve species. 'Midwinter', winner of an F.C.C., is extra-early in bloom. The dwarf form is popular in Japan, but its purplish-rose flowers are not a colour which is widely admired in the West.

It seems strange that more selections with superior colours have not been made and distributed from this extremely valuable species, a trim and floriferous asset in any garden. Oddly, *R. dauricum* is probably best known in the United States as one of the parents of the hugely popular Mezzitt grex, P.J.M., of which there are now numerous named clones vastly superior in flower colour. In the British Isles the best known *R. dauricum* hybrid is probably 'Praecox', a cross with *R. ciliatum* made by Davies in 1860, and winner of the Award of Merit, bestowed by the Royal Horticultural Society, in 1926.

A third species, *R. sichotense*, shows a strong resemblance to *R. dauricum* var. *sempervirens*, and has – for the time being – been described as a variant to *R. dauricum* by Cullen, who is waiting to see more material.

R. mucronulatum Turcz. "with a small point"

yíng hóng dùjuān "welcome red rhododendron"

li

hiii

g viii

gii

k

kv

kiii

eii

hiii

d

ci

b

ei

a

SUBSECTION **SALUENENSIA** (Hutch.) Sleumer

(Series *Saluenense*)

DR. MARY FORREST

One hundred years ago, in 1894, the French missionary Abbé Soulié discovered one of the most delightful dwarf rhododendrons, *Rhododendron saluenense*. It is the type species of Subsection *Saluenensia*, which comprises two species and several subspecies as follows: *R. calostrotum*, Balf. f. and Kingdon-Ward; *R. calostrotum* ssp. *keleticum*; *R. calostrotum* ssp. *riparioides*; *R. calostrotum* ssp. *riparium*; *R. calostrotum* ssp. *calostrotum* and *R. saluenense* Franch. *R. saluenense* ssp. *saluenense* and *R. saluenense* ssp. *chameunum*.

These species are small evergreen shrubs to 1.5m (5ft) high, often with a prostrate or hummock-forming growth habit created by an intricate pattern of branches. The 1–3cm (⅖–1¼in) long leaves are clothed in dense crenulate scales. The flowers, held in terminal clusters of one to five blossoms, are funnel-shaped to rotate, surrounded by five calyx lobes. The flowers are downy on the outside and the ten stamens are noticeably hairy at the base. The magenta or purple flowers are borne from mid-spring to early summer. The large calyx and rotate flowers with a downy throat distinguish these species from other dwarf rhododendrons. The members of this subsection are native to India, western China, and Burma, where they occur at altitudes of 3000–4000m (9800–13,000ft).

Rhododendron saluenense derives its name from the River Salween, one of the three rivers of the region. Following the original introduction by Soulié in 1894, from a location near his French mission station at Tseku in the Yunnan region of China, later introductions were made by George Forrest in 1914 and 1921–22, by Frank Kingdon-Ward in 1926 and by Joseph Rock in 1923–24.

Rhododendron saluenense attains a height of 1.5m (5ft), but the subspecies *chameunum* is usually prostrate. The leaves are oblong, to 3cm (1¼in) long by 1.5cm (⅔in) wide, the upper surface elepidote or lepidote, and the lower surface densely covered with scales. Flowers are borne in terminal clusters of one to three magenta or purple flowers. In 1945, it was given an Award of Merit when exhibited from Exbury, England.

Two hybrids are recorded: *R*. 'Hipsal' (*R. saluenense* × *R. hippophaeoides*) and *R*. 'Prostsal' (*R. saluenense* × *R. prostratum*), which were raised by E. J. P. Magor of Lamellen, Cornwall, in 1926.

They are hardy shrubs, suitable for use in a rockery or peat bed. These species can be cultivated in sunny situations, but frost pockets must be avoided. Little or no pruning is required. They are free-flowering and give an annual display of blossoms. When not in bloom, the low-growing, dense habit forms a useful ornamental barrier between a path and a shrubbery.

Rhododendron calostrotum 'Gigha', which was given a First Class Certificate in 1971 when exhibited by Peter Cox, is freely available in the trade; the availability of the other species is less frequent.

In Ireland, members of this subsection are cultivated at John F. Kennedy Park, Co. Wexford; Malahide Castle, Co. Dublin; Mount Congreve, Co. Waterford; Mount Usher, Co. Wicklow and Glenveagh National Park, Co. Donegal. In these gardens they are grown as single specimens or in massed plantings of rhododendrons.

While they look attractive in Western gardens, my reason for selecting these species is based on a photograph. Some years ago I was given a Chinese rhododendron book which contains some marvellous photographs of rhododendrons in their natural habitat. A centrepiece illustrates a flock of sheep amongst an expanse of pink shrubs, with snow-capped mountains behind. They were grazing amongst *Rhododendron saluenense* ssp. *chameunum*.

R. saluenense Franch. (from the Salween river, China) nù jiāng dùjuān "Nu Jiang rhododendron"

a

gv

b

ci

ei

gii

d

k

kiii

kvi

kvii

hiii

e

SUBSECTION **SCABRIFOLIA** (Hutch.) Sleumer

(Series Scabrifolium)

BEVERLEY LEAR

*I*f you are the sort of person that likes the big, blowsy flowers that rhododendrons are well known for, then this is not the subsection for you. If on the other hand, you favour the unusual, the delicate and subtle pleasures of small, exquisitely formed flowers, or if you are just mad keen on rhododendrons and have a spot where apparently none will grow, then there is surely a plant within *Scabrifolia* for you. Indeed, Lionel de Rothschild wrote in his account (1934) of plants in his newly created rock garden at Exbury that: *"Rhododendron racemosum, hemitrichotum* and *pubescens* grow where nothing else will."

Subsection *Scabrifolia* is currently recognized as consisting of six species and three varietal forms, although some dissension is apparent as to the taxonomic status of *R. spinuliferum* var. *spiciferum*, which continues to be regarded as a full species by some authorities. I suspect that long may such arguments rage. Broadly speaking, the members of Subsection *Scabrifolia* (*R. racemosum, R. hemitrichotum, R. mollicomum, R. pubescens, R. scabrifolium* and *R. spinuliferum*) are a clearly related group with similarities in flower form, habit and mode of growth. They are amenable shrubs, which are characteristically rather small (usually 50cm–3m/20in–10ft tall), exhibiting some likeness to species in the larger Subsection *Triflora*, from which they are distinguished by bearing their flowers in the leaf axils only. There are two species that I wish to mention more fully: *R. racemosum* and *R. spinuliferum*.

By far the most common member of the subsection in cultivation is *R. racemosum*, which is often seen with distinct, rich red branchlets that stand in admirable contrast to the rather small dark grey-green, rounded leaves that are glaucous beneath. The floral display is refined, with flowers (usually a deep rose-pink) somewhat less than 2.5cm (1in) long, borne over a long main season and with some tendency to be recurrent later in the year. In the wild, *R. racemosum* grows throughout Yunnan and South-West Sichuan and, significantly for the horticulturalist, it favours dry hillsides, including limestone outcrops.

The species was first discovered by Abbé Delavay in April 1884 on He-Chan mountain, Yunnan, and first cultivated at the Jardin des Plantes in 1889 whence it was introduced to Kew in the same year. Since that time, it has been reintroduced many times and consequently appears in our gardens in several guises. For example, plants introduced by Forrest (F19404) are favoured for

their free-flowering and dwarf habit, while plants introduced by Rock (R59717) from the Lichiang Range are reputedly much hardier than the Forrest plants. *R. racemosum* 'Rock Rose' is a tall-growing clone with good red stems and strong pink flowers, which received an Award of Merit (A.M.) in 1970, and *R. r.* 'White Lace', also a tall form but with white flowers, was awarded an A.M. in 1974.

Rhododendron spinuliferum is my real favourite. This plant, with its softly hairy leaves and its tendency, shared by others in the subsection, to throw rather long and lax, upright "watershoots", may not be everybody's choice of a fine garden plant, but I love its small, upright, urn-shaped flowers with protruding style and stamen (a unique form within the genus) in yellow, orange, peach, brick-red or crimson. Appearing in mid- and late spring, they are to me as little parrots, which hover about the bush in flashy colours (often a blend of several colours arranged together in stripes). Strangely, it was never collected by Messrs. Forrest, Kingdon-Ward or Rock, but was once again first found by Abbé Delavay in July 1891 in southern Yunnan.

Despite its apparently unique form, *R. spinuliferum* crosses readily with other species and there has been some interest among breeders to improve hardiness while conserving the unusual flower form. The Exbury Stud Books list no fewer than 18 different crosses made using *R. spinuliferum* during the 1930s and indicate the popularity of the species during that period. Sadly, named forms such as 'Quetta' and 'Berylline' are apparently now lost from cultivation, but other good hybrids remain including: 'Crossbill' (*R. spinuliferum* × *R. lutescens*) with tubular, yellow flowers raised by both Lionel de Rothschild and Mr. J. C. Williams of Caerhays; 'Seta', a distinct and excellent hybrid with upright, tubular flowers of light pink with a conspicuous, deep pink stripe bred by Lord Aberconway 'Razorbill', which is similar to 'Seta' but of greater hardiness, from Mr. Cox at Glendoick.

Clockwise from top: *R. scabrifolium* var. *spiciferum* Franch. máo yè dùjuān "hairy-leaved rhododendron," "bearing spikes"; *R. scabrifolium* var. *scabrifolium* Franch. cāo yè dùjuān "rough-leaved rhododendron," "with rough leaves"; *R. racemosum* Franch. yè huā dùjuān "axillary-flower rhododendron"

h*iii*

k ei

k*v*

c*i*

c

c*i* h*iii* k k*iii*

a

a

a

b

b

d

ei

h*iii*

g*ii*

k*iii*

c*i*

k*v*

Subsection **TEPHROPEPLA** (Hutch.) Sleumer

(Series *Boothii*, Subseries *Tephropeplum*)

LESLEY EATON

*M*any and varied are my favourites amongst the rhododendron species, so when I was asked to select just one subsection to write about from the list of illustrations for this book, I found the task a difficult one.

I finally chose *Rhododendron tephropeplum* in Subsection *Tephropepla*, as I had fallen for its charm some 25 years ago in my early days of "discovering" the species. I was examining the blooms on the benches at a local flower show when one of the judges, who noted I was taking a keen interest in the species, pointed out *R. tephropeplum* to me, turned the exhibit around and drew my attention to its eared calyx and the ash-grey covering on the leaves. On further investigation I then discovered that it was from these attributes that the name was derived.

My first plant of this lovely small rhododendron was the larger-flowered and -leafed form that is now known as the Arunachal Pradesh form, previously known as *R. deleiense*. This form prospers in my garden at Mount Dandenong, some 40 kilometres (25 miles) east of Melbourne, with its mild climate. It has continued to entrance me, not only with its reliably good flowering, but also with its purple new growth and its flaking, cinnamon bark. I later acquired the smaller-leaved form, which, like its larger-leaved sister, thrives in south-eastern Victoria and Tasmania.

Reginald Farrer, in May 1920, was the first plant explorer to discover *R. tephropeplum* growing at Chawchi Pass in upper North-West Burma. It was first introduced into cultivation by George Forrest in 1921, after an expedition to eastern Tibet. In 1930, Frank Kingdon-Ward collected *R. tephropeplum* from the Delei Valley in Assam. It was known at that time as *R. deleiense*, but in 1948 this plant was made synonymous with *R. tephropeplum*.

Rhododendron tephropeplum is found growing at altitudes of 2440–4300m (8000–14,000ft). Its native habitat covers a wide area from East Arunachal Pradesh, South-East Xizang (Tibet), North-East Burma to North-West Yunnan. There it is found growing not only on crags, rocks and cliffs, but also in screes and meadows and sometimes on limestone. Its growth varies considerably in habit and height. It can be quite bushy to upright or sprawling. My 25-year-old, larger-leaved plant has reached 1m (3½ft), with open growth, while the small-leaved form is only about two-thirds the size and is more fastigiate in habit. The *R.H.S. Species Handbook* states that *R. tephropeplum* can attain 1.8m (6ft). Growth habit is dependent on situation – straggly under shelter and more compact in open situations. Leaf shape and size and flower size and colour can also vary considerably. The flowers, which are tubular-campanulate, are in trusses of three to nine florets. Although the colour range can be from pale rose right through the scale to crimson-purple, white-flowered forms have been found, although these are quite rare. The form one sees mostly in our Australian gardens is the one with quite bright pink flowers in the red-purple group (73A in the R.H.S. colour chart). There are two features common to both forms: the lower surface of the leaves is densely covered with small black or brown scales and the corolla has a large, leafy, erect or spreading calyx. Flowering time is mid- to late spring, which is October to November here in the southern hemisphere (April to May in the northern hemisphere). This is the peak flowering time when a splash of pink is needed among the blues, mauves and yellows of many of the smaller rhododendrons.

Other members of Subsection *Tephropepla* must not be overlooked either. One such species is *R. auritum* – a soft lemon-flowered charmer with a rosy blush, where once again the flaky, cinnamon bark is an added attraction. *R. auritum* grows tall, sometimes reaching 3m (10ft). Its native habitat is the Tsangpo Gorge, South-East Xizang, where it grows at altitudes of 2100–2600m (7000–8500ft). This is another underrated plant, which is often neglected for its more flamboyant cousins.

Rhododendron hanceanum, added to this subsection by Cullen and Chamberlain, is also worthy of a place in the garden. A low, compact grower, it has creamy flowers and a most attractive bronze new growth which always catches the eye. As is the case with most of the species, it can be quite variable.

The final member of the subsection, *R. xanthostephanum*, was collected by Abbé Delavay in Yunnan. In the superior forms, it is a pretty, free-flowering species with bright yellow flowers. Once again it has the attractive bark that is a common feature of the subsection. Perhaps it is even better known as a parent of the lovely, older hybrid, 'Saffron Queen' – a hybrid which is still seen extensively – and of 'Lemon Mist', 'Meadow Gold' and 'Prairie Gold', to name a few newer American hybrids, which, as they are becoming more available, are more widely seen in our gardens.

R. tephropeplum Balf. f. et Farr.

huī bèi dùjuān "grey-clad rhododendron"

a

hvi

kv

k

b

gii

hv

g

ci

ei

d

Subsection **TRICHOCLADA** (Hutch.) Cullen

(Series *Trichocladum*)

Margaret Tapley

Subsection *Trichoclada* is a relatively small group of low-growing to medium-sized shrubs, none higher than 60cm–2.5m (2–8ft), and the lowest growing 30cm (12in) in height. The subsection has quite variable characteristics. There are both evergreen and deciduous species contained within it, and it shows no particular similarity to any other subsection. Its countries of origin are the north-west, west and mid-west of the province of Yunnan in China, the east and north-east of Upper Burma, and the east and south-east of Xizang (Tibet).

The flowers are carried at the ends of the branches or sometimes axillary in the uppermost one to three leaves. They are most commonly variations of yellow shades, including greeny-yellow, sulphur-yellow, pale-lemon-yellow or creamy-white – but can vary in some of the species within the subsection to reddish-yellow, reddish and sometimes, although rarely, there can be greenish or crimson spots on the posterior side. The corolla is bell- or widely funnel-shaped, and is 1.3–2.8cm (½–1⅛in) long.

The common characteristic by which members of Subsection *Trichoclada* can be identified with certainty is the fact that they possess vesicular-type scales on the lower surface of the leaves, as distinct from the more usual entire-type scales.

It is interesting to note the names of the different species in this subsection and their countries of origin. *Rhododendron caesium*, which is deciduous, was once thought to belong to Subsection *Triflora*, but since 1963 has been placed in Subsection *Trichoclada*, because of its vesicular scales. *R. rubrolineatum* (evergreen or semi-evergreen), *R. trichocladum* (deciduous), and *R. lepidostylum* (evergreen), which is the illustrated example, all come from the province of Yunnan in the south of China. *R. lithophilum* and *R. melinanthum*, both deciduous species, come from Burma, and *R. mekongense* (deciduous) and *R. viridescens* (evergreen) come from Xizang. *R. rubroluteum* was introduced by Kingdon-Ward after his expedition to China, Xizang and Upper Burma in 1922. Unfortunately, he neglected to tell us exactly where he found this unusual evergreen species with its bluish-green leaves and reddish-yellow flowers with crimson spots on the posterior side. It makes an interesting addition to any species collection as it is hardy, late-flowering and on top of all this has very good, blue young growth on its glaucous foliage.

The native home of *R. lepidostylum* is a relatively restricted area in West and Mid-West Yunnan. We are indebted for its discovery to the Scot George Forrest, one of the intrepid plant-hunters who risked life and limb to collect seed of new varieties of plants in China, Burma, Xizang and other Asian countries during the nineteenth and twentieth centuries. In 1919, Forrest discovered the plant growing on cliffs in open situations, on rocks where humus had collected, and in crevices high up on cliffs, or in ravines on the summit of the Jangtzow Shan, Shweli/Salween divide. It grows at elevations of 3050–3660m (10,000–12,000ft). Forrest introduced it in 1924 (No. 24633).

The flowers are widely funnel-shaped, pale yellow or yellow, and they are often produced in clusters of three. The plant is dwarf-growing, from 30cm to 1.2m (1–4ft). The soft yellow of the flowers is one of the most useful and adaptable colours to use in garden planning adding light and definition to a colour scheme and although the flowers are small, they have the endearing quality of a "look, see" plant, one that deserves close attention rather than one which brazenly trumpets its presence from a distance.

The leaves of *R. lepidostylum* are a striking, bristly, blue-green, and are extremely attractive in early summer when the new growth appears. They also take on a wonderful blue-red-purple shade in winter when the frosts touch them. Leaf-size is relatively small, 2–4.3cm (¾–1¾in) long and 1–2cm (⅕–¾in) broad, and of a regular, close-packed, oblong-oval shape. *R. lepidostylum* was given an Award of Merit for foliage in Great Britain in 1969, when shown by Capt. Collingwood Ingram at Benenden, Kent. Another valuable characteristic of the foliage is that it is aromatic.

I can find no record of any notable hybrids using this species as a parent. Perhaps this is a challenge for the future. The name "lepidostylum" means "with scaly styles", and it is a robust grower that will increase readily from cuttings. It is fairly easy to grow, given the usual rhododendron requirements of good drainage, adequate water, some shelter from wind and an acid soil.

R. lepidostylum Franch. "with scaly styles"
cāo máo dùjuān "coarse-haired rhododendron"

a

kv

hvi

hiii

d

ei

eii

ci

b

kiii

f

gi

gii

kvii

Subsection **TRIFLORA** (Hutch.) Sleumer

(Series *Triflorum*)

I VOR T. STOKES

S ome of the most garden-worthy species of rhododendron are to be found amongst members of Subsection *Triflora*, and I am surprised that they are not seen more often in cultivation.

There are almost 20 species within the subsection and between them they encompass a range of colour that is possibly unequalled by any other group within the genus. Forms of *Rhododendron concinnum* can be a rich glowing reddish-purple; a deep primrose-yellow is found in *R. lutescens*; the lilacs, lavenders, pinks and whites in *R. oreotrephes, R. davidsonianum, R. yunnanense* and their kin; while perhaps the nearest to a true blue in any rhododendron is to be seen in the best clones of *R. augustinii*. Even within the species there can be a wide range in colour, not just of the flowers themselves, but also in the spotting or flare on their upper petals and occasionally in the colour of their stamens. So great can this variability be that some forms are inevitably considered more desirable than others. It is therefore wise to look for named clones of species, especially where they have received awards.

As the name of the subsection suggests, the flowers are not borne in large trusses, yet such is their size relative to the leaves that, when in full bloom, they can completely obscure the foliage. When seen from a distance, a single bush flowering in woodland can give the impression of a myriad resting butterflies. They are definitely among the most reliably floriferous species in the genus and it is indeed a very poor year when the 'Trifloras' do not put on a good show.

Rhododendron yunnanense, in all its forms, is a charming plant and as reliable as any in the wealth and beauty of its flowers. It occurs in every shade between rose-pink and white, while the spotting on its petals can be green, brown, orange or red. The specimen portrayed here is given an almost "orchid-like" appearance by the striking red spotting on a white background.

Although variable in stature, they are for the most part medium-sized shrubs, with a twigginess and smallness of leaf that can bring a feeling of airiness and grace to plantings, contrasting well with the often ponderous, larger-leaved rhododendrons. As they do not form a dense canopy of foliage, the filtered sunlight that reaches the ground allows them to be quite closely underplanted with choice herbaceous woodland plants, such as *Roscoea, Uvularia, Tiarella,* cyclamen and many others.

The pale-yellow-flowered *R. keiskei* is perhaps the most variable species in size and ranges from a shrub of more than 2m (6½ft) in height to dwarf, almost prostrate, plants. The latter forms, which include 'Yaku Fairy' and 'Ebino', are well suited to cultivation among acid-loving alpine plants and are a delightful addition to the peat garden.

I enjoy these plants for features other than their flowers. *R. triflorum,* upon which the subsection is based, does not rank with the finest for its flowers, which are of a yellow that easily gets lost among the leaves. As the plant matures, however, it develops a framework of shining, mahogany-coloured branches from which the bark unfurls in papery sheets. If space can be found for it, then perhaps it should be grown. I find it an arresting sight when it is caught in the slanting rays of an early morning sun. Scent is found in *R. lutescens,* but as it is an early flowerer, its blooms can frequently be lost in frosty weather. However, they tend to open in succession over a long period and there are normally some that escape damage. The new growth of this species is most attractive: the young, willow-shaped leaves have a bronze coloration that slowly becomes green as the season advances.

Although the best forms of the species can hardly be improved upon, breeders have created some quite exquisite hybrids between the 'Trifloras' and species from other lepidote subsections. *R. oreotrephes, R. yunnanense, R. rigidum* and three or four others in this subsection have been successfully married to that equally graceful group the 'Cinnabarinas'. Plants in Subsection *Cinnabarina* possess characteristic, pendulous flowers and these crossings have produced a race of offspring with a subtle beauty that combines the better attributes of each parent. Interesting hybrids have also resulted from crosses with some of the dwarf lepidote subsections, creating plants suitable for the rock garden.

With a wide distribution over the mountainous areas of China, Burma and the eastern Himalayas (plus *R. keiskei* from Japan), the species occur in a variety of situations: as isolated plants in light woodland, as single-species thickets or mixed with other shrubs such as *Berberis, Buddleia* and *Rosa,* sometimes in quite exposed locations. They seem to be more tolerant of drier soils than many rhododendrons and thus are well suited to average garden conditions, where they will always attract an admiring eye.

R. yunnanense Franch. (from Yunnan)
yún nán dùjuān "Yunnan rhododendron"

a

b

k

k*iv*

k*v*

k*vii*

g*vii*

m*i*

h*iii*

e*i*

c*i*

o

g*ii*

t

d

g

SUBSECTION **UNIFLORA** (Hutch.) Sleumer
(Series *Uniflorum*)

ISOBYL F. LA CROIX

Subsection *Uniflora*, a small group of small plants, is now considered to include four species, one of them having two varieties. Although well known in cultivation, they are apparently rare in the wild – only one is at all widespread. This is *R. pumilum*, which is known from India, Nepal, Bhutan, North-East Burma and Xizang (Tibet). *R. pemakoense* occurs in Xizang and Assam near the Xizang border, but *R. uniflorum* var. *uniflorum*, *R. ludlowii* and *R. uniflorum* var. *imperator* have each been collected only once, the first two in Xizang and the last in North-East Bhutan. The status of some of the species in the subsection could change if more wild collections became available; cultivated material is unreliable from a botanical point of view because of the risk of hybridization. Sadly, although some areas of China have opened up in recent years, Xizang has not and it may be some time yet before the final word is said, in as much as it ever is in such matters.

All these species are dwarf, sometimes prostrate, shrublets with flowers large for the size of plant and freely borne. The smallest is *R. pumilum*, Kingdon-Ward's 'Pink Baby', which rarely grows much taller than 10–15cm (4–6in) and has pink, bell-shaped flowers about 2cm (¾in) long. The other species have funnel-shaped rather than bell-shaped flowers; the most distinctive is the yellow-flowered *R. ludlowii*. I have never grown this species – it is supposed to be difficult and not a "good doer". However, it has proved to be a good parent and has produced some splendid dwarf, yellow-flowered hybrids, notably *R.* 'Chikor' (*R. chryseum* × *R. ludlowii*) and *R.* 'Curlew' (*R. ludlowii* × *R. fletcherianum*). We grow both of these and the latter, in particular, is a superb plant.

The remaining species have pink, mauve-pink or purple flowers and are rather similar. *R. uniflorum*, in both its varieties, has slightly smaller, more wide-open and usually deeper-coloured flowers than *R. pemakoense*, with scales scattered over the lower surface of the leaf rather than set closely together. Var. *imperator* is quite prostrate, while var. *uniflorum* tends to form a mound. Both varieties flower in mid- to late spring.

The species illustrated here, *R. pemakoense*, is one for which I have a great affection, as it is one of the first rhododendrons we grew, and we have had it in all our non-tropical gardens. Many people like to know the derivation of a plant's name: *pemakoense* comes from the type locality Pemakochung in southern Xizang, where it grows in the Tsangpo Gorge at an altitude of 3000m (10,000ft). A neat plant, usually forming a low mound about 30cm (12in) high, it is said to spread sometimes by stolons, although we have not experienced this. It has small, dark green leaves and five-lobed, funnel-shaped, mauve-pink flowers up to 3.5cm (1½in) long, which are produced in such profusion as to cover the plant. Unfortunately, it does have one disadvantage: it flowers early, in early to mid-spring, and the buds are therefore very susceptible to frost. When we grew it in Surrey, England, where mild days in late winter would lure plants into false expectations of spring, it was all too frequent that we found the depressing sight of fat buds promising flowers turned to brown.

Various measures can be taken to try to prevent this. Frost pockets should, of course, be avoided – a raised bed can help. Part of a garden where other plants tend to come into flower late, such as the base of a north wall, can be successful, and possibly a covering of horticultural fleece, which was not available in those days, might help.

Here, on the exposed north-west coast of Scotland, where large rhododendrons are impossible to cultivate until windbreaks are fully and successfully established, these dwarf species come into their own. A slab of stone placed on edge gives good protection and although spring may be late, it tends to come without false starts and *R. pemakoense* can successfully smother itself with its pink funnels.

Subsection *Virgata*, see page 109.

R. pemakoense Ward
jiá dān huā dùjuān "false solitary-flower rhododendron"

d

b

ci

kv

ei

f

hvi

gii

hv

kiii

k

a

Section **POGONANTHUM** G. Don

(Series *Anthopogon*)

PETER COX

*E*ver since I started to grow rhododendrons 40 years ago, I have been fascinated by the plants of the *Pogonanthum* section, previously known as the *Anthopogon* series and, at the time when *The Species of Rhododendron* was published in 1930, also known as the *Cephalanthum* series. These species are unlike all other rhododendrons, having a totally different aroma from the foliage. This aroma has been likened to that of pineapple, thuja, Friar's Balsam and bog myrtle. It varies somewhat from species to species, my favourite being that of *R. kongboense*, which has a peculiar strawberry-like odour, and I often rub the foliage in passing. The characteristic overlapping lacerate (irregularly edged) scales may give the leaf undersurface a rich reddish-brown colour. Their stature varies from tight and prostrate to tall and straggly, up to 2m (6½ft), while their leaves are 0.9–6cm (⅓–2⅓in) long, and linear to oval. The most attractive small flowers have a tubular corolla with spreading lobes, giving an overall effect like a daphne. The colour is very variable, even within some of the species, ranging from pure white through shades of pink to deep strawberry-red, and yellow. Their texture is often like limp tissue-paper, which makes them easily spoilt by heavy rain. While some will flower within two to three years from a cutting, others take ten years or more. The flowering time is between early spring and early summer.

These species have a widespread distribution in the wild, from the Afghanistan-Pakistan frontier eastwards to North Sichuan and South Sichuan in China. They are found at widely varying altitudes from 2400 to 5500m (8000–18,000ft), in open forest representing almost the last shrubs before vegetation ceases, only to be beaten in the genus *Rhododendron* by *R. nivale* var. *nivale* at 5900m (19,000ft).

All the species grow best in areas with relatively cool summers such as Scotland, with many being tricky to cultivate in southern England. Very good drainage is essential. All are sensitive to fertilizer, and prefer a less acid soil than suits most rhododendrons, rich in organic matter.

The best species for general cultivation is *R. trichostomum*, which is often found in the wild growing on dry banks. The best forms have either clear pink or pure white flowers. *R. anthopogon* ssp. *anthopogon* is usually one of the slowest to bloom but has attractive pale yellow to deep rose-coloured flowers. This species is widespread and plentiful in the Himalayas where the leaves are frequently collected by the local Buddhists, who dry them and use them for incense. *R. cephalanthum* is perhaps my favourite in the section. I saw splendid, compact specimens covered with flowers growing on cliffs on the Mekong/Salween Divide, North-West Yunnan, in spring 1994. The lovely, almost prostrate-growing Crebreflorum Group has beautiful pink flowers, which appear on young plants. The larger-leaved var. *platyphyllum* was lost to cultivation for a number of years but was reintroduced by my son Kenneth and myself recently. *R. collettianum*, with the largest leaves in the section, has cream-coloured flowers to match in size. *R. kongboense* grows upright but tidily with its unusual, deep strawberry-pink flowers early in the spring. *R. sargentianum* is very neat and compact with dark, shiny leaves. The pale yellow flowers almost hide the leaves in a good season; 'Whitebait' (Award of Merit [A.M.] 1966) has slightly larger leaves than the type, cream-coloured flowers and is generally easier to grow. *R. primuliflorum* is the species I have seen most of in the wild. The tall forms are possibly the least desirable plants in the section with most taking years to start flowering; one rewarding exception is 'Doker La' (A.M. 1980), a marvellous clone which smothers itself in clear pink flowers every mid- or late spring.

Further treasures remain uncollected in the wild with the following notable examples. The tiny, hairy-leaved *R. pogonophyllum* in Central Bhutan, *R. fragrans* in Central Asia, *R. anthopogonoides* (lost from cultivation) in Gansu, China, and *R. rufescens* in Sichuan, China. In 1990 I was prevented from finding this latter species by an unexpected, early-autumn snowfall.

Clockwise from top:

R. kongboense Hutch. (from the province of Kongbo, China);

R. trichostomum Franch. máo zuǐ dùjuān "hairy-mouthed rhododendron";

R. sargentianum 'Whitebait' (after C. S. Sargent, d. 1927, a former Director of the Arnold Arboretum) Rehd. et Wils. shuǐ xiān dùjuān "narcissus rhododendron"

a

a

gii

k

kiii

k

kv

k

kiii

kv

ci

ei

k

gii

b

b

ci

gi

e

hiii

gii

d

b

ci

hiii

f

Section **VIREYA**

(Section *Vireya*, Series *Pseudovireya*)

GRAHAM SMITH

The *Vireya* section of genus *Rhododendron* is the most diverse of all. It is further divided into seven subsections and one of these, *Euvireya*, into a further seven series. All of these divisions are based on scale characteristics – often, extremely different-looking plants are placed in the same subsection.

Vireyas are subtropical, mountainous plants occurring mainly through Malaysia, Borneo and Papua New Guinea, with some outliers to be found in Asia and northern Queensland, Australia. As they inhabit regions close to the Equator, they are equal day-and-night plants and have no defined seasons. Rainfall is more likely to trigger growth and flower cycles, and a speciality of some species is that this can happen several times in the year. Others follow the usual pattern of one flowering a year, often from midwinter, when the season is naturally dry, to early spring.

In cultivation, the plants generally follow the same pattern, which makes them excellent conservatory plants or garden plants in the warmer countries. They are basically frost-tender, but many species tolerate a few degrees of frost without damage.

The flowers come in an extraordinary array of shapes and sizes, from tiny bells, through slender, tubed beauties to enormous trumpets, the largest in the genus; the colours are more vibrant than those of their Asiatic cousins and are often distinctly bi-coloured. Many have powerful scents that will fill a glasshouse or garden.

No plant better illustrates the diversity of *Vireyas* than *R. christii*. It comes from the Central Highlands of Papua New Guinea, where it occurs both terrestrially on mossy banks or rock faces or as an epiphyte. As long as the drainage is excellent it will grow well, showing distinct triangular foliage that occurs in whorls of three that clasp the stem. New growth is an attractive reddish-purple in colour, typical of many hot-climate plants. The growth can be lax or upright to 2m (6½ft).

The flowers of *R. christii* are extraordinary, being distinctly bi-coloured. The short, curved tube is bright yellow, while the flat, wide lobes are orange-red. They occur in trusses of three to six, dangling from the shoot tips, and flower for many weeks.

When collecting in Papua New Guinea, I found a typical *R. christii* growing on a wet, moss-covered log, its flowers hanging either side of the log like beacons. It grew in the base of *R. superbum*, another gem of a species. Stiffly upright with thick, scaly leaves, rusty-red when young, it has the most beautiful, soft-pink or white flowers, strongly scented of carnations. On some specimens seen in the wild these were up to 15cm (6in) across.

The most common species in Papua New Guinea is *R. macgregoriae*, which colonizes disturbed areas very readily. The flowers are small, star-like in shape and held in dense clusters. Orange is the prime colour, but bright yellows through to reds can be found. Although normally a compact plant, it can grow to 5m (16ft). It is excellent in cultivation and for hybridizing.

Two of the most vibrant species, both with trumpet-shaped flowers, are the brilliant yellow *R. laetum* and orange-red *R. zoelleri*. These are tall, open shrubs with handsome foliage, but it is as parents that they excel and they are also repeat-flowering.

Rhododendron commonae is a compact plant which grows at relatively high altitudes and is therefore hardier than some. Growing naturally on the margins of forests, quite often only 1–2m (3–6½ft) from frost grassland, the typical form has deep-red, almost tubular flowers. They are good plants in cultivation.

At the opposite end of the size scale are two red species: *R. saxifragoides* and *R. rubineiflorum*. The former is a very specialized plant, growing on a few mountain tops in glacial bogs. It forms a low cushion of narrow foliage and anchors itself to the base rock with a long taproot. Surviving frost, wind, occasional snow and intense light radiation, it produces single red flowers that stand atop the foliage – alas, it is a challenge to grow and flower!

Rhododendron rubineiflorum is much more amenable and grows and flowers quite easily if given a peat block or tree-fern log to grow on, just as it does in the wild. Tiny, round leaves on thread-like stems sprawl out from the plant centre, forming a dense mat. Comparatively large red bells in ones and twos dot the plant for many months, making it a jewel among rhododendrons.

Rhododendron javanicum var. *brookeanum* comes from Borneo and has one of the most spectacular of flower trusses. In the best forms they rival any rhododendron, but it is the vibrant reddish-orange colour that sets them apart.

These are but a few of a large number of species in the *Vireya* section which are becoming popular garden plants in the warmer areas of the world. Although they have been in cultivation for over 100 years, we still know little about them and, for me, they are the most exciting challenge in the rhododendron world!

R. christii Foerster (after H. Christ, 1833–1933, Swiss botanist)

k

a

c

ei

hiii

gii

b

b

SUBSECTION **AFGHANICA** Cullen

(Series *Triflorum*, Subseries *Hanceanum*)

MARIANNA KNELLER

*P*reviously placed in Series *Triflorum*, subseries *Hanceanum*, Dr. Cullen has given this species its own subsection.

This is a very poisonous, low-growing, semi-prostrate shrub, which attains a height of 50cm (20in) or so. The flowers can range from creamy-yellow, whitish-green to white. The leaves are dark olive-green, bluish tinged above with a pale green underleaf.

Originally discovered by Dr. J. E. T. Aitchison in 1879 on his travels to the Kurran Valley on the Afghanistan–Pakistan border, it grows at 2135–2900m (7000–9500ft) on the cliffs and ledges in the limestone rocks of the Abies Forest. It was introduced to several gardens in Britain but became lost to cultivation in the 1940s. It was re-introduced by Hedge and Wendelbo in 1969, after they had found it in the Alishang Valley in the province of Laghman, where the Nomads protect it from their grazing sheep.

Flowers mid- to late spring.

R. afghanicum Aitch. et Hemsl. (from Afghanistan)

SUBSECTION **FRAGARIIFLORA** Cullen

(Series *Saluenense*)

MARIANNA KNELLER

*P*reviously placed in the *Saluenense* series, Dr. Cullen in his revision has given *R. fragariiflorum* its own subsection. It is between Subsection *Saluenensia* and Subsection *Campylogyna*.

It is a very low-growing shrub, forming carpets of slightly fragrant, pink-flowered tussocks of just 15cm (6in) in height. The flowers vary from crushed strawberry-red to dark plum-purple.

It was first found by Kingdon-Ward in June 1924 (No. 5734) at Temo la, southern Tibet, growing at 4600m (15,000ft) on exposed rocky hillsides. He noted that it could also be seen growing all over the Alpine regions of South Tibet. *R. fragariiflorum* is also quite happy growing in swampy pastures or open, sloping turf-lands. It was rediscovered in 1947 by Ludlow, Sheriff and Elliot (No. 15828) in South-East Tibet and Bhutan.

R. fragariiflorum is a pretty-flowered, hardy little shrub and, although quite rare in cultivation, would be suitable for the rock garden, where it would grow happily alongside *R. camtschaticum*.

Flowers late spring to early summer.

R. fragariiflorum Ward "with strawberry-(coloured) flowers"

Subsection **GENESTIERIANA** (Hutch.) Sleumer

(Glaucophyllum Series)

MARIANNA KNELLER

Originally placed in the *Glaucophyllum* series by Balfour, Dr. Cullen has given this species its own subsection. It is a highly distinctive species, quite unique with its highly glaucous, plum-like bloom on its deep port-wine-coloured flowers, calyx and capsule. The flowers are similar to *R. campylogynum*.

It is a free-growing, upright shrub, with decorative purplish-plum or reddish-brown branches, and it can reach a height of 4m (12ft) or more. The new growth is quite a decorative, attractive reddish-purple colour, turning to bright green leaves, which have a white glaucous bloom to their underleaf.

It was named after Père Genestier, a French missionary and friend in Tibet to George Forrest, who discovered this species (No. 17824) in April 1919 in North-East Upper Burma.

It is a very tender species, only growing successfully in Britain in the milder, well-sheltered gardens of Cornwall and at Brodick Castle Gardens, Isle of Arran.

Flowers mid- to late spring.

R. genestierianum Forr.

huî bái dùjuān "grey-white rhododendron"

Subsection **LEDUM** (Hutch) Sleumer

(Genus *Lepidotum*)

MARIANNA KNELLER

This is a recent union between the genus *Ledum* and the genus *Rhododendron*. Until 1990, when Kron & Judd published the results of their research, *Ledum* was considered a separate genus. Kron & Judd have placed it in Subgenus *Rhododendron* in its own Subsection. It is possibly a sister group to *Edgworthia*. Seven species share this new subsection. These are very hardy, pretty, compact to upright-erect evergreen shrubs, up to 2m (6ft) high. The leaves are oval to linear and often aromatic; the young shoots are often covered with white- to rust-coloured woolly tomentum. The flowers are white and carried in compact terminal clusters, from late spring to early summer. *R. neoglandulosum*, *R. columbianum*, *R. groenlandicum* and *R. subarcticum* are from North America and Greenland, and the latter can also be found in northern Asia along with *R. tolmachevii*. *R. hypoleucum* comes from eastern Asia and *R. tomentosum* from Europe and northern Asia. They all favour peaty places and open pine forests.

R. ledum

SUBSECTION **LEPIDOTA** (Hutch.) Sleumer

(Series *Lepidotum*)

MARIANNA KNELLER

*T*his subsection contains some very pretty, delicate-flowering shrubs, growing to 1.8m (6ft) or so. The three species that form this group are *Rhododendron lepidotum, R. lowndesii* and *R. cowanianum. R. lepidotum* is very attractive, with delicate, flat-faced flowers in an umbel, rarely over three in number, in various hues of yellow, pink, purplish-red and white. The name "lepidotum" means "beset with scales", which is appropriate as its underleaf is completely covered with overlapping scales. J. D. Hooker collected it from the Sikkim Himalayas in 1848, though it had featured earlier in Wallich's catalogue 758 and in Royle's *Illustrated Botany of the Himalayas* in 1839. It has a wide area of distribution (Nepal, Bhutan, North-East Burma and China) and is a variable species. It is quite hardy, although it favours a sunny, sheltered area, and it is also a suitable candidate for a rock garden. *R. lepidotum* flowers in early summer.

Rhododendron cowanianum is a fairly recent discovery collected by O. Polunin in 1949 in central Nepal, where it grows in forest clearings, and on scree slopes and rocks. Originally placed in the *Trichocladum* series, because of its deciduous leaves, it is more closely allied to *R. lepidotum*. It grows to 2.5m (8ft) and the flowers are pink, purple and wine.

Rhododendron lowndesii is a creeping shrublet up to 30cm (1ft) high with flowers of pale yellow spotted with yellow ochre. It was discovered by Colonel D. G. Lowndes in 1950 in Nepal, where it grows in rock crevices, cliff ledges and peaty banks at 3050–4600m (10,000–15,000ft). It is a difficult, slow-growing species that needs protection in the winter months. Flowers late spring to early summer.

The specimen drawn here of *R. lepidotum* was supplied by Mr. K. Rushforth from his collection.

R. lepidotum Wall. ex G. Don

lín xiàn dùjuān "scaly-glandular rhododendron"

SUBSECTION **MONANTHA** Cullen

(*Uniflorum* Series)

MARIANNA KNELLER

*T*his is a new subsection, containing four species previously placed in the old series *Triflorum, Uniflorum* and *Boothii*.

These species are rarely found in cultivation. They are epiphytic or free-growing shrubs and in this respect are related to Subsections *Maddenia* and *Boothia*.

The main type species is *R. monanthum*, which is quite widespread in habitat. It is a spreading shrub and quite straggly. It loves the shady places and margins of the pine forests or the rocky slopes among the scrub and is at home on the cliffs of South-East Xizang (formerly Tibet) and Yunnan, where it was first collected by George Forrest in the Mekong-Salwin Divide. It is found at 2740–4270m (9000–14,000ft). It is epiphytic, loving moss-clad trees and shrubs. The flower is solitary, campanulate and bright yellow. The outside is scaly, with leaves glaucous on the undersurface, and densely scaly. *Rhododendron. flavantherum, R. kasoense* and *R. concinnoides* are very like *R. monanthum*.

R. monanthum Balf. f. et W. W. Sm.

yîduǒ huā dùjuān "one-flowered rhododendron"

SUBSECTION **VIRGATA** (Hutch.) Cullen

(*Virgatum* Series)

MARIANNA KNELLER

*R*hododendron virgatum, together with its ssp. *oleifolium*, make a very small subsection. It was an early introduction, first being found by J. D. Hooker in 1850 in the Lachen Valley, Sikkim, Himalaya. It is extremely tender and needs very carefully planned site conditions if it is to survive at all.

Peter Cox, in his excellent book *The Smaller Rhododendrons*, recommends "planting it on a bank where it can sprawl and be prepared to take cuttings regularly in case of loss".

A very floriferous species with flowers – one or two only – of a pale to deep pinky-mauve tones, it bears the flowers unlike most species in axillary rather than terminal buds. Its name means "with willowy twigs".

The specimen drawn here came from Wakehurst Place in Sussex, England, where it grows as a shrub.

Subspecies *oleifolium* ("olive-like foliage") is the Yunnan representative of *R. virgatum*, collected and described by the Abbé Delavay in 1886. A leggy shrub 60–90cm (2–3ft) high, it has pink to white flowers smaller than those of *R. virgatum*.

R. virgatum Hooker f.

liǔ tiáo dùjuān "willow-branch rhododendron"

PART III
AZALEA-TYPE RHODODENDRONS

When Linnaeus published his book on botanical classification in 1753, only nine rhododendron species were known to him; these included azaleas which he considered different enough to give them their own genus. This established a trend and to many people, azaleas are still not considered to be rhododendrons. Although many plant nursery catalogues give azaleas their own category, the taxonomists believe that azaleas are part of the great rhododendron genus. Azaleas are usually deciduous, or semi-evergreen, and their leaves often turn to lovely autumnal colours before falling.

The species placed in the four sections, which form Subgenus *Pentanthera*, originate mainly from North America, with the exception of one European species and two from Asia. Their flowers are white, through a range of yellow tones, to orange, vermilion and scarlet; they are often highly scented. These species are largish shrubs, growing up to 4.5m (15ft) or more.

Subgenus *Tsutsusi* has approximately 80 species in two sections; they are exclusively Asiatic, mainly from Japan. They are the hardier, mainly semi-evergreen species, mostly low-growing in habit. Their flowers are white, pink, red or purple-magenta – the anthocyanin colours – and they are devoid of any yellow pigmentation. I found that the vibrance of the magenta tones in some of the species was best represented by coloured inks, rather than watercolours, to obtain a true colour match.

Subgenus *Azaleastrum* with its Section *Choniastrum* has species that are in between azaleas and rhododendrons. A number of them are evergreen. The species in Section *Choniastrum* are tender and more suitable for the greenhouse. The lovely species *R. championiae* was collected from the Temperate House, at The Royal Botanic Gardens, Kew, and although carefully protected on its journey to the studio, reacted immediately to the change of temperature and never fully recovered its form before completely collapsing within a few hours. However, even in that short time, it imparted its uniquely delicate and beautiful character.

Subgenera *Mumeazalea* and *Candidastrum* each have a single species, both highly distinctive, while Subgenus *Therorhodion* with its three species is still considered by some taxonomists and horticulturalists to be a separate genus.

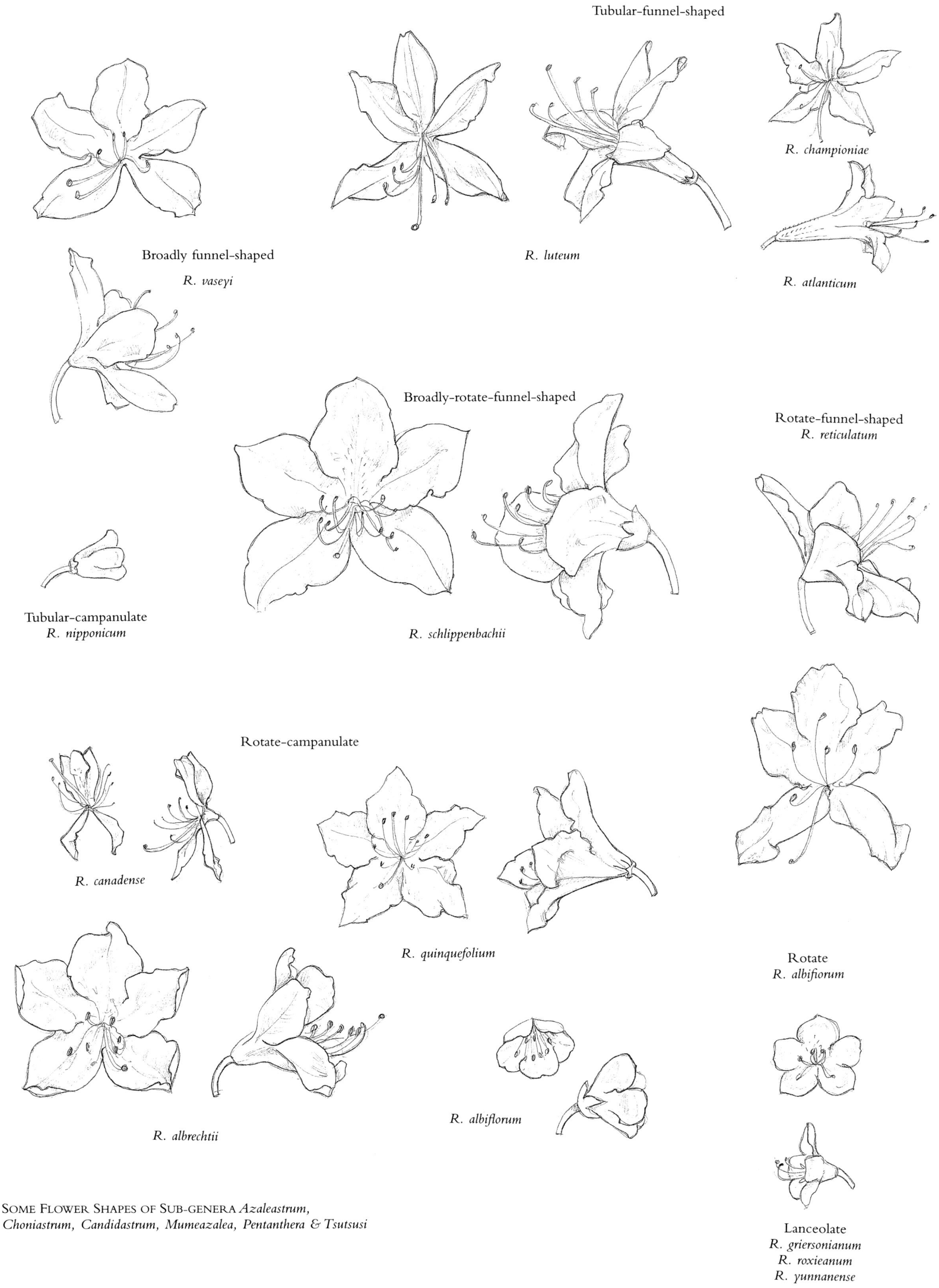

Tubular–funnel–shaped

R. championiae

Broadly funnel–shaped

R. vaseyi

R. luteum

R. atlanticum

Broadly–rotate–funnel–shaped

Rotate–funnel–shaped
R. reticulatum

Tubular–campanulate
R. nipponicum

R. schlippenbachii

Rotate–campanulate

R. canadense

R. quinquefolium

Rotate
R. albiflorum

R. albrechtii

R. albiflorum

SOME FLOWER SHAPES OF SUB-GENERA *Azaleastrum,*
Choniastrum, Candidastrum, Mumeazalea, Pentanthera & Tsutsusi

Lanceolate
R. griersonianum
R. roxieanum
R. yunnanense

Subgenus PENTANTHERA

(Section *Pentanthera*, Series *Azalea*, Subseries *Luteum*)

Steve Hootman

The long-cultivated and infamous pontic azalea, *Rhododendron luteum*, is well known to most growers of rhododendrons as a fragrant, yellow-flowered deciduous azalea with brilliant autumn foliage. Some also know it as a common and sometimes weedy poisonous plant, toxic to animals and humans alike. The role of this species in history is perhaps the richest and most ancient of any rhododendron, reaching back at least 2400 years.

Rhododendron luteum was first introduced into cultivation in Europe in 1792 by Peter Simon Pallas of Germany, who collected it in the Caucasus Mountains. These mountains and adjacent Turkey constitute the principle range of this species. There are also scattered and isolated populations in eastern Europe and western regions of the former Soviet Union. It occurs along streams and in swampy areas, on open wooded slopes, and on the shores of the Black Sea from sea level to around 2300m (7500ft).

Hardy to around -32°C (-25°F) and adaptable to a wide range of soils and growing conditions, the pontic azalea is widely cultivated and has even become naturalized in some areas. This species is cherished for its ease of cultivation and masses of fragrant, bright yellow flowers that appear unfailingly each year in early to late spring. In the autumn the foliage changes to scarlet, orange or yellow before dropping late in the season. *R. luteum* was widely used as a parent in many of the hybrid azalea groups.

The pontic azalea is a member of Section *Pentanthera*, which includes 15 to 18 species of deciduous shrubs native to Europe, Asia and the United States. All are in cultivation with several having been so for centuries. The species in this section typically have showy, often fragrant flowers of white, pink, red, orange, yellow or lavender and all shades in between. All but four of these are native to the eastern United States, with the greatest concentration in the south-eastern region of this country. *R. luteum* is the eastern Europe/western Asia representative. The other three outlying species are *R. molle*, which is native to China, *R. japonicum* from Japan, and *R. occidentale*, which hails from the western United States. Further notable species in Section *Pentanthera* are *R. periclymenoides*, the pinxterbloom azalea; *R. viscosum*, the swamp azalea; *R. calendulaceum*, the flame azalea; *R. japonicum*, the Japanese azalea; *R. molle*, the Chinese azalea, and *R. occidentale*, the western azalea.

R. periclymenoides, *R. viscosum* and *R. calendulaceum* are all native to the eastern United States. The latter occurs in the Appalachian

Mountains and was first introduced by John Lyon in 1806. In mid-spring, its large non-fragrant flowers in brilliant shades of orange, scarlet or yellow more than justify the appellation "flame azalea". *R. periclymenoides* and *R. viscosum* both have large ranges covering much of the eastern half of the United States. The pinxterbloom azalea was first introduced by Peter Collinson in 1734 and is found in a wide variety of habitats, but is especially common in dry woodlands of oak and pine. It has tubular flowers appearing in early spring in shades of pink or lavender, sometimes ranging to white, and usually not fragrant. The swamp azalea was the first rhododendron native to the United States to be introduced into England. It is usually found in moist woodlands and swampy areas and blooms in late spring to early summer. The white tubular flowers, which have a strong spicy fragrance, are very attractive against the newly emerged foliage.

Rhododendron occidentale is native to Oregon and California, where it occurs along the coast and into the mountains in habitats ranging from bogs to wooded slopes, serpentine ridges and ocean bluffs. It was introduced by William Lobb, who sent seed to the Veitch Nursery in 1851. The flowers of this species are fragrant and variable in size and colour, ranging from white to pink and yellow, usually with variously coloured spots and stripes.

Rhododendron molle and *R. japonicum* were first introduced in 1823 and 1830 respectively. Both have probably been cultivated for centuries as ornamental and for medicinal purposes in their countries of origin. The Chinese azalea is widely distributed in China, where it occurs in various open habitats. The flowers of this species occur before or with the leaves in shades of yellow, orange or sometimes red, usually with green spotting in the throat. The Japanese azalea is native to Japan in a wide variety of open and woodland habitats. Flower colour is extremely variable, ranging from yellow to orange, red and even pink.

The aforementioned species all played a large part in the original breeding programme in Europe, including the Ghent, Knap Hill and Exbury hybrids. They are still being used as a genetic resource in hybrid breeding programmes both in Europe and the United States. Numerous named and award-winning selections of each species are available.

R. luteum Sweet "yellow"

Subgenus **PENTANTHERA**

(Section *Rhodora*, Series *Azalea*, Subseries *Canadense*)

STEVE HOOTMAN

Upon graduation from college, I entered the profession of horticulture as Curator of a botanical garden in the Washington D.C. area. During those five years on the eastern coast of the U.S.A., I spent a great deal of time studying the native flora of that region. I was especially interested in the wild rhododendrons and other *Ericaceae* which grew in profusion there. Two of my favourite species were (and still are) *Rhododendron vaseyi*, the pinkshell azalea, and *R. canadense*, the rhodora. With their uniquely shaped flowers and brilliantly coloured autumn foliage, these plants stood out among all the azaleas I was to come across. These encounters with some of the finest species of *Rhododendron* were what initially sparked my interest in the genus.

The rhodora, *R. canadense*, was first described in 1756 from a specimen collected in Canada and growing at the Botanical Gardens in Paris, France. It was first introduced into England by Joseph Banks in 1767. This species is the most northerly occurring of all North American azalea species. Its natural range extends from Labrador, Quebec and Newfoundland in Canada, south through the north-eastern states in the U.S.A. to northern New Jersey. It grows in moist, highly acidic areas such as bogs, wet meadows, swamps and moist woodlands at elevations ranging from sea level to the summits and upper slopes of some of the lower mountain ranges. The rhodora is a stoloniferous shrub usually under 1.2m (4ft) in height. The non-fragrant, two-lipped flowers appear in early spring before the foliage and are very distinct, with ten stamens and the lower lip divided into two separate and narrow lobes. Flower colour is typically lavender to purple or red-purple, or occasionally white. The deciduous foliage can be quite colourful in the autumn, showing tones of purple, rose or yellow. This species makes a wonderful addition to any garden or collection, but it is especially useful in areas that experience extreme cold since it is hardy to at least -32°C (-25°F).

The pinkshell azalea, *R. vaseyi*, was first discovered in 1878 by George S. Vasey, who was then botanist in charge of the United States National Herbarium. It was introduced into cultivation by the Arnold Arboretum of Jamaica Plain, Massachusetts, soon after 1880. This species is native to only six counties in the mountains of western North Carolina, at elevations of 900–1650m (3000–5400ft). In the very limited geographical area where this species occurs, it can be a common or even dominant underplanting shrub species. I have seen it forming a pink haze under the leafless canopy of deciduous or semi-deciduous mountain forests as far up the slope as one could see. It is common in rocky seepages, on road cuttings, and on large moss-covered boulders and cliffs where there is water seeping through the moss. This species also occurs in bogs, spruce forests, ravines and swamps. Colour ranges from the typical rose or pink to pinks so dark that they appear almost red. Pure white and white with coloured flecks are also found.

As is often the case with plants of a very limited natural range, *R. vaseyi* has proven to be very adaptable in cultivation. Hardy to -27°C (-16°F), this species also performs well in hot and humid conditions such as those prevalent in the south-eastern U.S.A. To see the delicately shaped, five-lobed flowers covering a shrub up to 4.7m (15ft) in height is an experience to anticipate. The flowers appear before the new emerging foliage in early spring, and it is not uncommon to find red, pink or green spots on the upper lobe or lobes. They are not fragrant and have five to seven stamens, in contrast to the ten one finds in *R. canadense*. The narrow, willow-like foliage of *R. vaseyi* is deciduous and, when growing in full sun, turns a brilliant fiery-red before dropping in the autumn.

Together, these two species comprise section *Rhodora*, formerly subseries *Canadense*, in the subgenus *Pentanthera*. Each is unique and so different from other rhododendrons and even each other, that they were initially placed in their own genera.

There are no recorded hybrids with *R. vaseyi*. This is not surprising, since it is one of the few rhododendrons that will not cross with any other species or hybrid. *R. canadense*, on the other hand, will cross with other species and hybrids, although this is not a common occurrence with a tetraploid. It was crossed with *R. japonicum* by G. Fraser of Vancouver Island, Canada, in 1912 to yield *R.* 'Fraseri' (syn. *R. × fraseri*) with 4cm (1½in) purplish-pink flowers. There are several named clones of *R. vaseyi*, including: 'White Find', a pure white selection, and 'Suva', a clone with red-purple flowers and a white throat that received the Award of Merit (A.M.) in 1969 when exhibited by Edmund de Rothschild. The species as a whole received the A.M. in 1927.

R. vaseyi Gray (after G. S. Vasey, 1822–1893, who discovered it in North Carolina in 1878)

a

g*i*

c*i*

c*i*

h*iii*

k

g*ii*   o   g   d

Subgenus **PENTANTHERA**

(Section *Sciadorhodion*, Series *Azalea*, Subseries *Schlippenbachii)*

(Section *Viscidula*, Series *Azalea*, Subseries *Nipponicum* [see p125])

Prof. Dr. Siegfried Sommer

*R*hododendron schlippenbachii Maxim. was named after Baron von Schlippenbach, a Russian navy officer and hobby botanist who discovered the species in 1853, on the eastern coast of Korea. Karl Johann Maximovicz, head botanist and curator at the St. Petersburg Botanical Gardens from 1852, described the new species in 1870.

Together with the related species *R. pentaphyllum* and *R. quinquefolium*, it belongs to the group of Sciadorhodions, or umbrella rhododendrons (Greek "sikadon" means "shading roof") – their leaves are arranged like an umbrella around the shoots. Another common feature these species share is that their flowers have 8–10 stamens. There are several ways of classifying this group – as subgenus, section or series.

Schlippenbach's rhododendron had to wait for nearly 50 years before it became more widely distributed in European gardens – even in 1912 the famous dendrologist Camillo Karl Schneider referred to it in his *Illustriertes Handbuch der Laubholzkunde* (Illustrated Handbook of Deciduous Trees) as "still rare in cultivation, but certainly hardy". Today, this entirely winter-hardy, summer-green species is offered in the catalogues of all the leading rhododendron nurseries.

Its flowers are flat, saucer-shaped and light pink marked with a reddish-brown and, at 8cm (3⅛in) in diameter, are relatively large. They open before the leaves, towards the end of spring. It is possible therefore that, in an unfavourable position, a late frost may carry off its floral splendour. The plants suffer no further damage from this, soon producing a fresh greening in order to flower again the following spring. *R. schlippenbachii* makes a fascinating shape in the spring garden – the tender elegance of its blossom, reminiscent of wafer-thin Chinese porcelain, and the light green of the budding foliage are cheering after the long, dark winter.

This is also what the seven Buddhist monks must have felt when they discovered a picturesque lake with many bays at the foot of the Diamond Mountains. They decided to interrupt their pilgrimage here for three days of meditation. In the woods surrounding this so-called Three-Days'-Rest Lake, the Samilpo, this species covers large areas, growing 1–2m (3½–6½ft) high underneath flat-crowned Japanese red pines (*Pinus densiflora*).

The native habitat of our species stretches from eastern China, Korea and eastern Manchuria up to the area of Vladivostok, where it reaches the northern limits of its homeland. It grows from the coast to higher mountainous regions of up to 200m (650ft) above sea level. It is not particularly choosy about its site – I found the species on acidic, humus-rich, weathered clay soils of gneiss and granite (pH 3.9 to 4.5, under the pine-needle layers) as well as on the coast of Wonsan in the Japanese ocean in humus-rich, sandy soil interspersed with shell fragments. Here the pH measured 6.3 to 6.9. This species also proved relatively tolerant of dryness and strong sunshine. Such – for a rhododendron – relatively wide ecological tolerance is remarkable, and tree nurseries quite rightly therefore recommend it for ordinary soils.

I also really like this species because, when it is not in flower, it is so easy to recognize by its foliage. The leaves are reverse egg-shaped, widest in the front part and rounded at the tip or blunt with a short thorny tip. They are usually arranged in groups of five at the end of the shoots, in a seemingly random way, and do not carry scales. Before dropping in the autumn they turn – depending on their situation – a strong yellow to a glowing, carmine red, and thus end their vegetative time with a second colour climax. All these excellent features make *R. schlippenbachii* especially valuable for the gardener. With its minimal growth – 30-year-old specimens are about 70cm (27in) high with a spread of 1.5m (5ft) – and with its sturdy, umbrella-like shape it is very suitable for planting in small gardens. In the garden it creates an excellent contrast with taller evergreens. *R. albrechtii*, which blossoms at the same time in a strong red, sets off the luminosity of its flowers. For underplanting, carpeting primulas, *Omphalodes verna*, *Waldsteinia geoides* and *W. ternata* have proved especially successful. This as yet uncrowned jewel from eastern Asia deserves a wider use in our gardens. It has no need of further work from breeders seeking to make improvements: it is perfectly beautiful as it is!

R. schlippenbachii Maxim. (after Baron Maxim von Schlippenbach, Naval Officer and traveller)

dà zi dùjuān "the flowers are similar in shape to the character t "dà"

k

a

s

b

gii

d

ci

ei

o

o

hiii

gv

Subgenus TSUTSUSI

Section *Brachycalyx*, Series *Azalea*, Subseries *Schlippenbachii*
Section *Tsutsusi*, Series *Azalea*, Subseries *Obtusum*

HIDEO SUZUKI

Subgenus *Tsutsusi* is made up of Section *Brachycalyx* and Section *Tsutsusi*. In Section *Brachycalyx* over 20 species and varieties are known, and although they are usually deciduous, some of the species of southern origin are likely to be semi-evergreen. They all have a set of three spade-shaped leaves on each apex of the branchlets. Most grow to about 3m (10ft) high and carry five-lobed, purplish-pink flowers in spring. It seems that all the species of this section are hardy in Japanese gardens.

Among gardeners in the West the name *R. reticulatum* represents all the other species of this section, and many of them appear not to be known to Western gardeners. But the azaleas of this section appeal so much to Japanese hobbyists that some are known to collect and grow species of this section alone.

In the West, however, *R. reticulatum* is the only species in *Brachycalyx* which is commonly grown. Its prevailing colour of some strong shade of purple is probably the main reason for its neglect by some rhododendron enthusiasts. For those who like the colour range the sight of a healthy plant in full flower is irresistible, its impact not easily equalled. In autumn the foliage often colours richly, and can be spectacular.

A few other species in the section can be found in good collections, including perhaps *R. amagianum*, *R. nudipes*, *R. sanctum*, *R. wadanum* and *R. weyrichii*. Some Western gardeners feel that warmer gardens and sunnier conditions are more likely to produce good growth and good flowering.

In the Section *Tsutsusi* over 60 species have so far been classified. They are mostly indigenous to China and Japan, with a few coming from neighbouring islands. Almost all species in the wild are evergreen or semi-evergreen, with five-lobed, pale brick-red flowers, and grow to 1–3m (3½–10ft) in height.

Rhododendron kaempferi, named after a Dutch merchant, Engelbert Kaempfer, is extremely abundant in Japan, inhabiting an area that stretches from the cold northern island of Hokkaido down to the warm southern island of Kyushu, a range of nearly 3000km (1900 miles) across the entire Japanese archipelago. This rhododendron species is found from the roadside level up to an elevation of 2000m (6600ft). As the wide distribution range indicates, this is one of the hardiest azaleas for cultivation in gardens. Thus, it may be worth considering Section *Tsutsusi*, a large group in the Subgenus *Tsutsusi*.

Here is a brief survey of the many interesting cultivars: 'Kohzujima' shapes naturally as dwarf and compact bonsai, 'Shikizaki' flowers twice a year in both spring and autumn, and 'Himeyama' flowers from spring through to autumn, changing colour in the process. Flower colours range from white to grey to pink or red, while flower shapes range from the 'Koshimino' hose-in-hose; 'Tachisen-e', double and tiny double; 'Kinshibe', ten reddish stamens without petals; 'Ginshibe', ten white stamens without petals; and 'Shideyama', polypetalized.

Another popular azalea in Japan is unquestionably the enormous group of *R. indicum* hybrids called 'Satsuki'. Literally thousands of different crosses of this species have been introduced into the horticultural world in Japan over a period of hundreds of years. It is no exaggeration to say that every household in this country has at least one or two potted 'Satsuki' azaleas.

R. reticulatum D. Don ex G. Don "netted" (referring to the venation)

Subgenus AZALEASTRUM

Section *Azaleastrum*, Series *Ovatum* (see page 122)

Section *Choniastrum*, Series *Stamineum*

Mark Jury

Subgenus *Azaleastrum* contains Sections *Azaleastrum* and *Choniastrum*. During recent years taxonomists have been working on this subgenus to bring the classification up to date. Many existing specific names represent findings which are little known; many are probably not in cultivation. In his book *The Rhododendron Species*, Volume III (1992), page 316, Mr. H. H. Davidian treats *R. westlandii* as synonymous with *R. moulmainense*.

The *Azaleastrum* and *Choniastrum* groups hold a special place in our garden. These plants look like a link between azaleas and rhododendrons. The flowers within the subgenus are very similar in shape and colour, and close in form to a *Rhododendron augustinii* or *R. yunnanense*. While some perform better than others in our conditions, all seem to be very healthy and disease resistant.

When my father, Felix Jury, began his garden at Tikorangi, on the west coast of the North Island of New Zealand, he had a backdrop of 70-year-old *Pinus radiata* and our native podocarp, the rimu, planted by his grandfather. In his search for interesting material, Felix imported many plants, mostly sourced from England. Over the years, certain rhododendron varieties have made themselves at home, while others, particularly those requiring cooler conditions, have struggled and disappeared. One of the groups which has succeeded is the azalea-like plants from the *Azaleastrum* and *Choniastrum* sections.

Section *Azaleastrum* is represented in our garden by *R. bachii, R. leptothrium* and *R. ovatum* itself. In the early 1950s, the New Zealand Rhododendron Association was involved with importing species. On a visit to the As's Nursery at Massey, my father acquired a plant of *R. bachii*. It was the only one available, but where it came from is not recorded. This plant has thrived. It is now a perfectly healthy large shrub at 3m (10ft) high and 2.5m (8ft) wide – probably the largest one in cultivation as it is described in the 1980 RHS species handbook as not being in cultivation at all! It flowers profusely in late spring with 4cm (1⅝in) pale lilac-pink blooms borne in each leaf axil. The rounded flower shape seems to be typical of this group and there are small maroon dots at the flower centre.

Rhododendron leptothrium has been in the garden for 20 years, but again its source is unknown. It flowers at the same time as *R. bachii* but lacks the compact growth habit. This may be due to its being in a less favourable spot, as it is nevertheless healthy. The leaves are longer at 6cm (2⅓in) and the flowers are slightly larger and star-shaped. The lilac-pink is deeper and the spotting more rosy.

Our most recent addition to Section *Azaleastrum* is *R. ovatum* itself, purchased ten years ago from Pukeiti Rhododendron Trust, who had imported it from Edinburgh. It is now 1.5m (5ft) high, being both upright and bushy and is a real charmer. It has a smaller and rounder leaf. Last spring it covered itself with delicate flowers, almost identical to *R. bachii* but flowering earlier, and with a beautiful, delicate perfume.

The *Choniastrum* section is represented in our garden by *R. moulmainense* and *R. ellipticum*. Once again the early plant of *R. moulmainense* was from the New Zealand Rhododendron Association. The plant is in a cool spot and has struggled; its pink flowers are only seen occasionally. However, a plant growing in New Plymouth flowers profusely every year. The flowers are typically similar to the *Azaleastrum* section, lavender-pink in colour but a shade deeper and with slightly darker green-brown spotting.

Rhododendron ellipticum came to us recently from Pukeiti and has started flowering profusely this year as a 1.3m (4½ft) shrub. It has the same lovely fragrance as *R. ovatum*, but again has a bigger flower and is a shade deeper. As with the other varieties, it has a dozen or so flowers per branch tip. It is the last of this subgenus to flower for us, not showing its colour until very late spring.

Overall there is a great deal of similarity between the *Azaleastrum* and *Choniastrum* representatives in our garden. They have a special quality of refined delicacy and are excellent, healthy plants in our temperate and sometimes humid conditions.

R. championiae Hooker (after Mrs. Champion, the wife of its discoverer, J. G. Champion, 1815–1854)

SUBGENUS **MUMEAZALEA**

(Series *Semibarbatum*)

HERBERT SPADY

*R*hododendron semibarbatum has been belittled as "more curious than ornamental and best grown in someone else's garden" due to its paucity of floral display. Indeed, its yellowish-white small flowers, arising singly from lateral buds, are obscured by the newly opened terminal leaves. The flowers may be tinged with pink and have a few red spots. The character of the stamens is diagnostic of the species: the three lower ones are glabrous and about twice as long as the two hairy upper ones. It certainly would not qualify for a dramatic welcoming site in any garden in the flowering season, but despite its understated floral performance this rhododendron is not without interest.

I was first attracted to the plant on seeing its bright glossy green leaves unfolding early in the season at the Rhododendron Species Foundation. The leaves reached about 5cm (2in) in length and were about half as wide. They were elliptic, but coming to a wedge at the base and a point on the end, hence appearing to be basically pointed on each end. They were papery thin. The leaves were tinted with mahogany-red on the margins, a harbinger of colour to come. In the autumn the leaves turned bright yellow, orange and crimson before falling.

Its plant habit lives up to its homeland in mountainous southern and central Japan. It grows open with spreading slender branches, giving a very oriental appearance.

It is indeed curious and unique, being alone in Subgenus *Mumeazalea*. It is interesting from a genetic and evolutionary viewpoint, leading one to consider a number of unsolved mysteries. For example, just where does it fit in the genus and where does it fit in rhododendron evolution? Are there any hybrids of it or has anyone ever tried to hybridize it? Obviously there are more questions than answers about *R. semibarbatum*.

It is not a difficult plant to grow in most temperate rhododendron climates. Its place in the garden is best where its foliage qualities can be displayed without any thought to flowering. It is because of its foliage and "curious" nature that it has a place in my garden.

Subsection *Candidastrum*, see page 125.

R. semibarbatum Maxim. "partially bearded"

a

s

ci

ei

ei

k

gviii

ci

b

d

ei

hv

hvi

SECTION **AZALEASTRUM**

Section *Azaleastrum (Ovatum Series)*

MARIANNA KNELLER

A group of five species, related to azaleas, the rhododendrons in Section *Azaleastrum* have an affinity with those of Section *Choniastrum* (*Stamineum* series) with which they share the subgenus.

Both sections have flowers that are axillary, borne singly, although the species in section *Azaleastrum* are generally smaller and have only five stamens to their shorter corolla.

They are small erect shrubs with slender twiggy branches; the leaves are evergreen.

Rhododendron bachii is a 1.8m (6ft) high, twiggy shrub, with rosy-lilac flowers. Intermediate in character and geographical range between *R. ovatum* of eastern China and *R. leptothrium* of Yunnan, it has the leaf of the former and the calyx of the latter. First discovered by Léveille, it grows in the Hupeh, Kiangsi and Kweichow provinces of China.

Rhododendron hongkongense "from Hongkong" is a very pretty species, with white flowers speckled with violet. It flowers freely in its natural habitat on rocks of the Black Mountain, Hong Kong. It is a very tender species, which flowers in early spring. The specimen drawn here was supplied by the Royal Botanic Garden, Edinburgh.

Rhododendron leptothrium "with thin leaves" has bright green, handsome lanceolate leaves, and flowers of deep magenta-rose, sometimes with olive-green markings. It can be grown in sheltered, mild gardens. Its natural habitat is West Yunnan, where it grows at 2135–3355m (7000–11,000ft). It flowers from early to mid-spring.

Rhododendron ovatum "egg-shaped" can grow to 4m (12ft) high. The leaves are broadly ovate, the flowers white, lilac or pink, spotted pink or with a yellow blotch. It was first found by Robert Fortune on Chusan Island and eastern China. *R. ovatum* is a shy bloomer, tender, and flowers early to mid-spring.

Rhododendron vialii (after Père Vial) is rare in cultivation. It is from South Yunnan, where it grows at 1200–1800m (4000–6000ft). The flowers are crimson or pink, sometimes lilac. It has long-stalked obovate leaves.

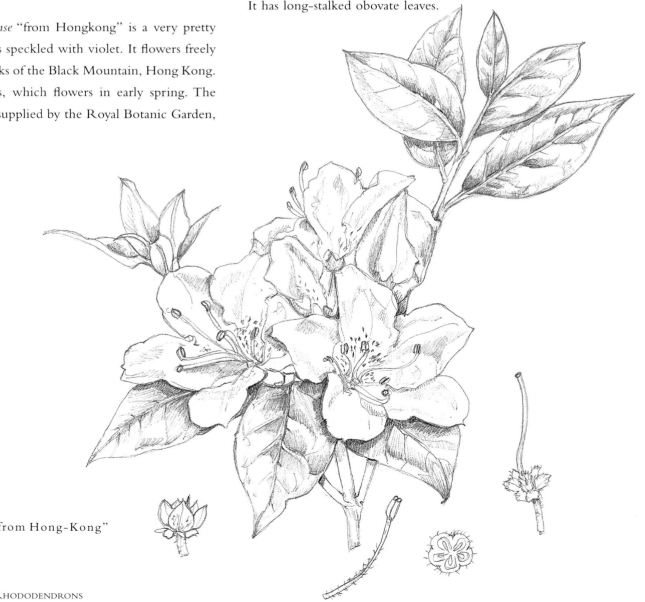

R. hongkongense Hutch. "from Hong-Kong"

Subgenus **CANDIDASTRUM**

(*Albiflorum* Series)

Marianna Kneller

A rather distinct form of rhododendron species, *Rhododendron albiflorum* "white flowers" is a native of the Rocky Mountains in North West United States, ranging from Oregon to British Columbia and eastwards to Colorado. It could be related to Section *Azaleastrum*, Series *Ovatum*.

It is a deciduous shrub, growing to 1.8m (6ft), with fastigate branches of a reddish-brown colour. The small, rotate-bell-shaped, white flowers, occasionally with yellow spots, tuck their ten stamens and pistil well into the corolla, which sits in a very large calyx. These flowers grow along the branches, well under the bright-green leaves that are clustered at the end of short branchlets; both the midrib under the leaf and the petiole have appressed brown hairs.

Millais mentioned *R.albiflorum* in his book *Rhododendron Species and the Various Hybrids* stating that "immense and almost impenetrable thickets of this species grow in some parts along and above the timber line of the Rocky Mountains at heights of 1200–3000m (4000–10,000ft)".

It is difficult to cultivate but is hardy.

R. albiflorum Hooker "with white flowers"

Section **VISCIDULA**

(Series *Azalea*, Subseries *Nipponicum*)

Marianna Kneller

T his section contains only this unusual species *Rhododendron nipponicum* ("from Japan"), first noted by Matsumura in *The Tokyo Botanical Magazine* vol. 13, 17, (1899).

It is a little-known plant, which originates from the mountains of Honshu in central Japan where it grows between 900 and 1400m (3000–4500ft) in the forests and on hillsides.

The six to fifteen white flowers are very small, 2cm (¾in) long 1cm (⅓in) broad. They have distinctive tubular, bell-shaped corollas, quite unique for azaleas, on slender pedicels, which nod in the breeze. The flowers tend to hide shyly amongst the leaves, which are similar to *R. schlippenbachii*. Being deciduous, the autumnal foliage produces a display of brilliant orange and crimson colours. Height up to 1.8m (6ft).

Rare in cultivation. Flowers early to midsummer.

R. nipponicum Matsum "from Japan"

Subgenus THERORHODION

(Series *Camtschaticum*)

Dr. Lothar Heft

The name *Rhododendron camtschaticum* Pall. 1784 indicates the place – Kamchatka – where Peter Simon Pallas (1741–1811) first discovered this rhododendron on his journeys through Siberia from 1771 to 1773, and the year 1784 when he described it.

At first glance, *R. camtschaticum* scarcely appears like a rhododendron at all, and some botanists, for example Bean (1980), took the view that it should be classified as a separate genus, *Therorhodion*. More recent research, however, has shown no significant differences from the genus *Rhododendron*, and *R. camtschaticum* will therefore probably remain within the genus.

The species is at home in eastern Siberia, Kamchatka, the areas around the Sea of Okhots, the Kuril Islands up to Sakhalin and northern Japan. It arrived in North America, the Aleutians and Alaska via the Bering Strait. It can usually be found in coastal areas and often on islands and peninsulas in these regions. It grows in damp and swampy places in mountainous areas, but also in crevices and on mountain tops in sandy clay, on rocks in the tundra and, in the alpine zone, among *Pinus pumila*, alder and birch.

Rhododendron camtschaticum is a low-growing, cushion-shaped shrub, rarely exceeding 30cm (1ft) in height, with procumbent and upright, light reddish-brown branches. It spreads by means of underground suckers. The leaves are deciduous, sessile, obovate, or spathulate-obovate, 1.2–4.5cm ($\frac{1}{2}$–1$\frac{3}{4}$in) long and 0.5–2.5cm ($\frac{3}{16}$–1in) wide, glandular and bristly; the margins and the prominent veins on the underside are bristly. Its beautiful autumn coloration is remarkable, ranging from yellow, orange and red to brown. The flowers appear singly or in pairs (rarely in triplets) on long shoots. They are moderately or densely pubescent, with long-stalked glands. The colour of the flowers is crimson, marked with small reddish-brown spots. Rarely, in its native habitat and in cultivation, one encounters plants with pink, red or white flowers. The main flowering time is late spring to early summer, but the plant often generously repeat-flowers in late summer and early autumn. It is best propagated by seed, but division and propagation of softwood cuttings in spring can be successful.

I first encountered *R. camtschaticum* more than 30 years ago in a tree nursery in spring where a few unsold leafless plants looked rather inconspicuous and not very tempting. However, I bought one, and the proprietor explained that he would not be offering *R. camtschaticum* in the future as there was little demand…

We planted this rhododendron in peat in a shallow clay pot and placed it in the south-facing window of our kitchen. It seemed to like this spot – the moist kitchen air made it shoot luxuriously, and soon in late spring numerous flowers appeared above the light green leaves. After flowering, we transplanted our rhododendron into the garden. It was love at first sight when I saw the plant in flower but why I fell for a rhododendron that is unremarkable for most of the year – I just don't know!

Many years later I was able to fulfil my wish to grow *R. camtschaticum* in its ideal situation. In 1970, in the Rhododendron Park in Bremen, Germany, a rock garden was constructed that was to be dedicated to alpine rhododendron species and varieties. The base and core of the rock garden were old granite cobblestones, which make for good water drainage. Walls were built of old granite kerbstones, boulders of various sizes and peat sods in order to create terraced planting areas with a humus-rich sandy soil for the rhododendrons. Like all other alpine rhododendron species, *R. camtschaticum* loves, even needs, the proximity of granite stone (primitive rock). Individual plants should always be planted on top of, and in between, fist- to head-sized stones. Their roots seek out the cool stones and appreciate the humidity of condensation on and underneath them – if planted like this the plants will burst with health. Today, nearly a thousand *R. camtschaticum* plants grow in the Rhododendron Park in abundance, the oldest about 20 years old and over 50cm (20in) in diameter and 25cm (10in) high.

It is an ideal plant for the rockery, but rather susceptible to soil dryness. In dry soil the leaves shrivel up from the edges inward, and the appearance of the plant is severely impaired until it drops its leaves in the autumn. After a short period of drought, as soon as it has rained or been watered, it will normally send out shoots and flower again in late summer or early autumn. Planted in shade the plants do not perform and refuse to flower. In a damp place, even at high temperatures and in full sun, there are no problems.

Despite all its good features and its extreme winter hardiness, *R. camtschaticum* will probably remain a species for the connoisseur. You have to love it . . . and not just at first sight.

R. camtschaticum Pall. (from Kamchatka)

yún jiān dùjuān "in the clouds rhododendron"

RHODODENDRON FOLIAGE

FÉLICE BLAKE

Once the flowers are over, rhododendrons enter one of the most fascinating phases of their annual cycle. Day by day, new life unfolds as foliage takes on the dominant mantle, demonstrating its wonderful diversity, unrivalled by any other genus. For a major part of the year, rhododendrons are not in flower, and gardeners depend on them as foliage plants. With careful planning, a rhododendron collection can hold something of interest for a large part of the year.

I acquired my first rhododendron species, *Rhododendron griersonianum*, nearly 40 years ago, and it introduced me to the wonderful world of indumentum which is simply one of the many worthwhile attributes of rhododendron foliage.

Everyone who has suitable woodland conditions in a congenial climate should grow some of the big-leafed species in the *Grandia* and *Falconera* subsections – little does it matter that they may take many years to flower. In many of the species in these subsections the new foliage stands up like cockades of white kid enhanced by red bracts, with indumentum slowly attaining differing shades as the leaves mature to look as outstanding as the flowers that will ultimately bloom. Just to feel the thick, tawny indumentum on *R. falconeri* and its subspecies *eximium* is a delight.

Some species make outstanding lawn specimens, and these can be appreciated without companion plants vying for attention. In particular, various forms of *R. arboreum* such as ssp. *cinnamomeum* and the form we used to call ssp. *campbelliae* bear the hallmark of quality. Their shapely, dark green leaves, plastered with silver or woolly tawny to white indumentum on the undersides of their leaves, are a spectacle and more so when ruffled by the wind, or when the new growth unfurls. It does not take many years for these plants to reach their full grandeur, achieving a tree-like stature in our gardens as they gain in height and girth.

A well-known species is *R. yakushimanum* ssp. *yakushimanum* in the *Pontica* subsection, which is widely grown and has lovely thick suede indumentum. In my own garden I have five different forms, all worthwhile and making an interesting group with related plants, including the narrow-leafed ssp. *makinoi*, *R. smirnowii* with its dense, white indumentum and *R. japonicum* var. *japonicum* (*metternichii*) with its handsome, dark green leaves with compacted indumentum.

Subsections *Taliensia*, *Neriiflora*, *Maculifera*, *Campanulata*, *Fulva* and *Lanata* are awash with wonderful species sporting an amazing variety of luxuriantly prolific indumentum. Perhaps the most widely grown and treasured species in these subsections are the sumptuous *R. bureavii* and *R. mallotum*, the former with its softer leaf with rufous indumentum and the latter with slightly deeper indumentum and showing its prominent leaf veins. However, my personal choice is the incomparable *R. elegantulum*, and I would place this species in the very front rank among its kind. It unfurls its new growth in a lovely, light silvery-cinnamon indumentum, which deepens to a very rich pinkish shade before maturing to a deep cinnamon. One intriguing facet of this rhododendron is that it bears its leaves for three years and one can glory in the three different stages of colour of the indumentum on the same stem.

One must not overlook other *Taliensia* species such as *R. roxieanum*, particularly in the form with incredibly narrow, dark leaves, which has been likened to a porcupine! And now to the midget *Taliensia*, *R. proteoides*, a most desirable species and a fascinating one. My plant is still only a few inches high although I acquired it about eight years ago. One most outstanding species insofar as new foliage-colouring and indumentum are concerned must be *R. campanulatum* ssp. *aeruginosum* with its intensely metallic-blue new leaves with pure white indumentum which later turns cinnamon. At its brilliant spring best, it is guaranteed to stop visitors in their tracks.

Another favourite is *R. pachysanthum* in the *Maculifera* subsection with its new growth covered with white fur which persists for a long time, making a striking contrast to its companions. This is one of the comparatively recently introduced species. The *Fulva* subsection provides one of my most treasured rhododendrons: *R. fulvum*. This species has the marvellous advantage of holding its leaves upright throughout the year, except for a few short weeks in mid-winter, and thus displaying its beautiful, deep cinnamon indumentum for all to see. My plant, over 20 years old, is near to the verandah at the front of my house where I can enjoy it. When lit at night by the verandah light, it assumes an aura of mystery as the indumentum takes on an extra glow.

Subsection *Lanata* provides the charming *R. tsariense*. The very name invokes visions of far-away mountains stretching into the mist. It is a delightful species, rather small-leafed, with beautiful indumentum and elegant new growth, making a small, shapely shrub in my garden.

Subsection *Williamsiana* holds the distinctively lovely *R. william-*

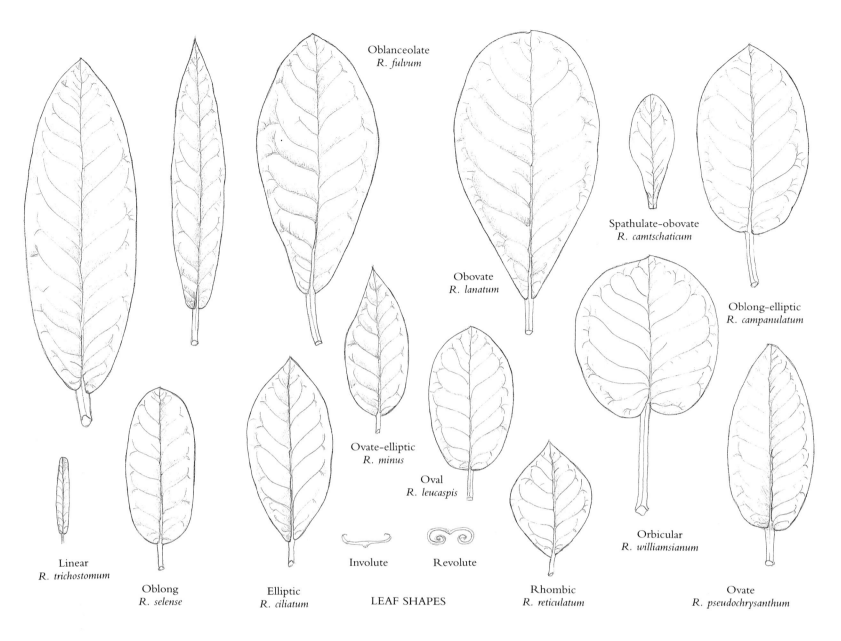

Oblanceolate
R. fulvum

Spathulate-obovate
R. camtschaticum

Obovate
R. lanatum

Oblong-elliptic
R. campanulatum

Ovate-elliptic
R. minus

Oval
R. leucaspis

Involute

Revolute

Orbicular
R. williamsianum

Linear
R. trichostomum

Oblong
R. selense

Elliptic
R. ciliatum

LEAF SHAPES

Rhombic
R. reticulatum

Ovate
R. pseudochrysanthum

sianum. With its small, rounded leaves and gleaming, bronzy new growth, it takes its place amongst the élite. Subsections *Fortunea* and *Thomsonia* also have many interesting foliage plants with attractive, rounded leaves. Among these the L. & S. form of *R. thomsonii* with its very glaucous new foliage remains attractive throughout the year, and so too does *R. orbiculare*.

Let us now look at the scaly lepidotes, and consider that distinctive enigma *R. edgeworthii*, which is so different to most other species in that it bears both indumentum *and* scales under its bullate leaves. The lepidotes are full of fascination, and Subsection *Maddenia* boasts some of the most magnificent, notably *R. nuttallii* with its large, strongly bullate leaves and purplish new growth, and not forgetting the little, golden, hairy-leafed *R. valentinianum*.

For those who delight in willowy shrubs, many of the species found in Subsection *Triflora* are attractive all the year around. *R. lutescens* charms with its graceful, bronzy, long-lasting new leaves, and the superlative *R. augustinii* always looks good with its variety of leaves – I must always feel the soft velvety leaves of the First Class Certificate form as I pass by! Surprises in the Triflorums must

be *R. zaleucum*, with its silver-backed leaves, the glaucous *R. oreotrephes* and the little *R. keiskei*.

Among the remaining lepidotes, a variety of delightful foliage characteristics are displayed. The most outstanding of the smaller species must be *R. lepidostylum* in Subsection *Lepidostyla* with its new blue leaves fringed with hairs. The blue-leafed *R. fastigiatum* in Subsection *Lapponica* and the glaucous-backed leaves in the *Campylogyna* and *Glauca* subsections provide more interest, and in particular many of the Lapponicas turn very bronzy in winter. Remember, too, the remarkable species in Section *Pogonanthum*.

When autumn arrives, we are blessed with an array of beautiful tints in many of the deciduous rhododendrons, including *R. albrechtii*, *R. schlippenbachii*, *R. luteum* and *R. occidentale*.

For those who delight in delving a little further into the sometimes hidden attributes of their species, I would suggest that they acquire a little, hand-held pocket microscope. To see the rather ordinary scales on *R. keiskei* leaves transformed to beautiful golden craters or the furry undersides of the leaves of *R. pendulum* transformed into a wild forest takes one into another world.

1a

2 *R. glischroides*

2a

1 *R. dictyotum* 'Katmandu'

3 *R. rex* ssp. *fictolacteum*

3a

1 *R. roxieanum*

1a

2a

2a

2 *R. eximium*

3 *R. haematodes*

3a

3b

1 *R. traillianum*

1a

4a

4 *R. trichanthum*

2a

3 *R. taliense*

2 *R. niveum*

3a

2b

5 *R. lutescens*

5a

1 *R. bureavii*

1a

2 *R. smithii*

3 *R. mallotum*

2a

3a

1 *R. wardii*

1a

3a

1b

2 *R. macabeanum*

2c

2a

3c

2b

3 *R. crassum*

3b

1 *R. watsonii*

2b

2 *R. griersonianum*

1a

3b

2a

3 *R. ponticum*

3a

1 *R. schlippenbachii*

1a

4 *R. reticulatum*

4a

2 *R. luteum*

3 *R. arborescens*

3a

5 *R. albrechtii*

5a

2a

1a

1b

2 *R. arboreum*

1 *R. cinnabarinum*

4a

4b

4 *R. orbiculare*

3 *R. pseudochrysanthum*

2a

3a

3c

3b

5b

6a

5 *R. williamsianum*

6 *R. yakushimanum*

6c

5c

5a

6b

1a

1 *R. decorum*

1b

3a

4a

3 *R. trichanthum*

4 *R. fulvum*

2 *R. oreotrephes*

2a

2b

5c

5b

5a

5 *R. glischroides*

CONTRIBUTORS

Dr. Björn Aldén has been the chief dendrologist and scientific curator at the Göteborg Botanical Garden, Sweden, since 1976. Part of his responsibilities is a public education programme using lectures, articles, exhibitions and radio and television programmes. In the early 1970s he made taxonomical and plant geographical studies on the alpine flora of Greece, and he is the co-author of *Mountain Flora of Greece*, vol. I, published by A. Strid, 1986. He has led, and participated in, several collecting expeditions to the Easter Islands, North-West Africa, China and the USA.

Bruce Archibold became interested in rhododendrons as a result of a visit to Sheffield Park Garden in Sussex, England, where the superb scent of *Rhododendron × loderi (R. fortunei × R. griffithianum)* made a lasting impression on him and his wife. As a result he joined the RHS Rhododendron and Camellia Group in the late 1970s, becoming Bulletin Editor and later, on the retirement of the Hon. Edward Boscawen as Chairman, taking on that office, which he has held ever since. He confesses to leaning to species plants, but he and his wife still grow *R. × loderi* in their five acres in Devon.

Warren Berg was a Boeing 747 captain who flew, until 1983, regular trips to Japan where he scouted the plants of the islands of Yakushima and Cheju. Since then he has made numerous trips to Sikkim, Bhutan, Xizang, Korea and South-West China. He has introduced many rhododendron hybrids and, in addition, has been a conduit for acquisition by the Rhododendron Species Foundation of countless numbers of the finest forms of rhododendron species from Great Britain, Japan and other parts of Asia. He is a former board member of the American Rhododendron Society and has been awarded the Gold Medal by them. Now, he is President of the Olympic Peninsula Chapter, holds a number of positions in the Rhododendron Species Foundation and owns one of the finest rhododendron species gardens in the USA. 'Patty Bee' and 'Ginny Gee' are perhaps the most successful hybrids he has developed, and one rhododendron species, *R. bergii*, was named in honour of his wife and himself.

Douglas Betteridge has worked at Exbury Gardens in Hampshire, England, for 40 years, the last 18 as Head Gardener. In 1994, the Royal Horticultural Society awarded him the Associateship of Honour – held by just 100 people at any one time – in recognition of his dedication and work at Exbury with the famous collection of Rothschild rhododendrons and azaleas. He was instrumental in providing Marianna with

the species she needed to paint the illustrations in this book. He will retire from the garden this year.

Félice Blake has been growing rhododendrons for nearly 40 years, and enjoys a one-acre garden at Kallista in the Dandenong Ranges east of Melbourne, Australia. Interest in the species rhododendrons began early. She belongs to a number of local and overseas horticultural societies and has served on the committees of the Australian Rhododendron Society, the Australian Lilium Society and the Ferny Creek Horticultural Society. She joined the American Rhododendron Society in 1978 and has contributed many articles to its journal. She has also written for other societies, and is author of a booklet on the George Tindale Memorial Garden. Over the years, Félice has given numerous presentations on rhododendrons, locally as well as to various Chapters of the American Society and to other societies in the Pacific North-West. With her late husband, she has experimented with rhododendron hybridization, and has done serious work on the hybridization of liliums – good rhododendron companions.

John D. Bond has been the Keeper of the Savill and Valley Gardens in Windsor Great Park in Windsor, England, since 1970 and a Council Member of the Royal Horticultural Society for the past nine years. He is also the Chairman of the RHS Rhododendron and Camellia Committee and a long-time rhododendron enthusiast.

Alan Clark is Curator of the gardens at Muncaster Castle, Ravenglass, Cumbria, England. The garden was started in the 1700s and contains many fine trees and a large collection of rhododendron species grown from seed collected by Ludlow, Sherriff, Forrest and Kingdon-Ward. Under his supervision, the 100-acre garden is undergoing complete restoration, including a 20-acre Sino-Himalayan glade. He also runs the Nursery and Plant Centre, which can normally offer at least 500 varieties of rhododendron.

Peter Clough started his horticultural career in the non-ericaceous environment of pH 7.8 as propagator at the Isle of Ely Horticultural Institute in Wisbech, Cambridgeshire, England. He was seduced by the lower pHs of the west coast of Scotland and moved to the Isle of Gigha in the Inner Hebrides in the mid-1960s. Here he had the enviable task of producing at least three plants of every rhododendron in the wonderful collection of Sir James Horlick at Achamore Gardens. Many of these plants were transferred to the Isle of Arran and now form the "Horlick

Collection" in the lovely garden of Brodick Castle. On the death of Malcolm Allan he took over as Head Gardener of Gigha and remained there until the end of the Horlick era. He then forsook rhododendrons for a ten-year sojourn in the Isles of Scilly, in charge of the exotics of Tresco Abbey Garden, but returned to the rhododendron fold later to take on Inverewe Garden's magnificent plant collection. Here he now exists happily at pH 4.5. In 1994 he at last was given the chance to experience rhododendrons in the wild at their epicentre in North-West Yunnan and hasn't stopped talking about it since, especially and ironically the species he found growing on magnesian limestone at – guess what? – pH 7.8.

Ambrose Congreve considers himself lucky to have witnessed the progress of a vast enterprise as it rapidly took shape: by the late 1920s, Mr. Lionel de Rothschild had achieved a great plant collection and massed colour effects at Exbury, a garden only begun in about 1919. At Mount Congreve, Waterford, Ireland, he has tried to emulate Mr. Lionel's concept of massed plantings without repetitions dotted around at random in bare earth. Ambrose Congreve was honoured with what he considers a somewhat undeserved Veitch Memorial Medal. Later, Her Majesty the Queen of Holland sent a Secretary of State, together with her Ambassador in Ireland, to present the Orange Order of Nassau to the Garden Director, Herman Dool, in recognition of his work at Mount Congreve since 1961.

Peter Cox is the Director of Glendoick Gardens Ltd, rhododendron nursery and garden centre in Tayside, Scotland. He is a member of the American Rhododendron Society (and was vice-president of the Scottish Chapter), a member of the Rhododendron Group, and member and honorary director of the Rhododendron Species Foundation. He is Chairman of the Annual Convention of the American Rhododendron Society to be held in Oban, Scotland, in 1996. He is the joint author of *Modern Rhododendrons, Modern Shrubs, Modern Trees, The Encyclopedia of Rhododendron Hybrids, Cox's Guide to Choosing Rhododendrons;* and author of *Dwarf Rhododendrons, The Larger Rhododendron Species, The Smaller Rhododendrons, Cultivation of Rhododendrons, Rhododendrons Wisley Handbook.* He has also written numerous articles for the *Journal of the American Rhododendron Society, The Rhododendron Yearbook, Country Life, The Plantsman* and *The Garden*. Peter Cox has raised many well-known and award-winning rhododendron hybrids, such as 'Curlew', 'Chikor', 'Ptarmigan', 'Razorbill', 'Panda' and 'Squirrel'. He has been actively involved in seven

expeditions to China, including the first British expedition after the Communist take-over in 1981, known as the Sino-British Expedition to China (SBEC), and expeditions to Turkey, Nepal, Bhutan and North-East India. He has been awarded the Victoria Medal of Honour, the Scottish Horticultural Medal, the Gold Medal of the American Rhododendron Society and the Loder Cup of the RHS.

Dr. Hugh R. Dingle is a retired anaesthetist living in Jersey, Channel Islands. As an enthusiastic amateur he has been growing rhododendrons for 30 years, and he is a member of the Rhododendron, Camellia and Magnolia Group of the Royal Horticultural Society. He has travelled to Nepal and Yunnan, where he was able to see rhododendrons growing in the wild.

Lesley Eaton is a retired school teacher. She has been a committee member of the Victorian Branch of the Australian Rhododendron Society since 1974 and is, at present, President of the National Council of the Society and President of the Victorian Branch. Through involvement with the Society and other horticultural bodies, she has developed her knowledge of the genus *Rhododendron*. In her three-acre garden in the Dandenong Ranges near Melbourne, Australia, she has over 800 different rhododendrons, both species and hybrids. Her garden has been open to the public as part of Australia's Open Garden Scheme for the past seven years.

Dr. Mary Forrest is lecturer in Horticulture at University College Dublin, Ireland. Her interest in rhododendrons developed during her tenure of the Heritage Gardens Fellowship when she prepared an inventory of exotic woody plant collections in Ireland, subsequently published as *Trees and Shrubs Cultivated in Ireland*. Prior to her present appointment, she was Head Gardener at Glenveagh National Park, Co. Donegal, the home of the late Henry McIlhenny. Mary is a member of the Rhododendron, Camellia and Magnolia Group of the Royal Horticultural Society, and is branch representative for Ireland.

David Goheen studied Chemistry at the University of Washington in Seattle, Washington, U.S.A., where, after serving in the Army during World War II, he obtained a PhD in organic chemistry and botany. He worked for over 30 years as a Research Scientist in the Central Research Laboratories of the Crown Zellerbach Corporation. His love affair with rhododendrons began in 1959. He has been a member of the American Rhododendron Society

(ARS) for 32 years, has served the Portland, Oregon, Chapter in many offices and, in 1988, received the Gold Medal of the ARS in Williamsburg, Virginia, for service in the genus *Rhododendron*. He was also the President of the Rhododendron Species Foundation in Federal Way, Washington, from 1983–4 and is still on their Board of Directors. He has travelled to most of the major rhododendron-growing areas in the world, including western China and Xizang. In his two-hectare garden in Camas, Washington, he grows over 200 species of rhododendron and thousands of hybrids, many of which are his own creations and introductions.

Harold E. Greer developed his interest in rhododendrons as a child in the 1950s, hybridizing his first rhododendrons at about age seven. He developed, owns and operates Greer Garden, a commercial nursery in Eugene, Oregon, U.S.A., which specializes in rhododendrons as well as unusual and rare trees and shrubs, sold by mail order. He is the author of *Greer's Guidebook to Available Rhododendrons*, and co-author of *Rhododendron Hybrids*. He has been director of the American Rhododendron Society for years and is now also the national president. The Society has awarded him both the Bronze and the Gold Medals for his work. He is a keen photographer and has had many thousands of his photographs published. He has also travelled widely and visited gardens and gardeners all over the world.

Alan Hardy was born and bred into a rhododendron garden. There followed a short flirtation with another genus, daffodils, until after an illness it was suggested to him by Mrs. Roza Stevenson that he devote himself to the genus *Rhododendron*. She also taught him all he knows about rhododendrons. For several years he edited the *RHS Rhododendron Yearbook*, and held the positions of Chairman and Vice-Chairman of the Rhododendron Group. He is the Vice-Chairman of the RHS Rhododendron Committee and Floral B, which also deals with all kinds of woodland plants. He raised several award-winning plants in the *Maddenia* series, and his family won both the McLaren and the Loder Cups.

Dr. Lothar Heft is a Consultant for the German Rhododendron Society and edited its Yearbook until 1992. He teaches in the faculty of Nursery Sciences at the Universities of Hanover and Hamburg, Germany. He studied at the Humboldt University and Technical University of Berlin, and researched into fertilization and propagation of rhododendrons at the Institute for Fruit and Nursery Sciences of the Technical University at Hanover. From 1967 to 1992 he was Director of the Botanical Gardens and the Rhododendron Park in Bremen, which he extended to contain the largest rhododendron collection on the European continent, including greenhouses for non-winter-hardy tropical and subtropical rhododen-

R. carolinianum Hutch. "from Carolina, U.S.A."

drons, and, since 1986, an Azalea Museum with a collection of *R. simsii* hybrids. From 1970 he was involved in the construction of a rock garden for alpine rhododendron species and cultivars. He has written on rhododendrons and azaleas and is the co-author of *Rhododendron und immergrüne Laubgehölze* (Rhododendrons and evergreen broad-leaved trees). He travelled to Nepal to see rhododendrons in their native habitat, and to England and Scotland to learn about rhododendrons and to update the Bremen collection.

Steve Hootman has been interested in plants all his life. He studied Botany at the University of Montana and graduated with a Bachelor of Science in Public Horticulture from Purdue University in 1987. From 1987 until 1992 he was Curator/ Horticulturalist of the Winkler Botanical Preserve in Alexandria, Virginia. Since then he has been working for the Rhododendron Species Foundation in Washington, U.S.A., where he currently is the Curator.

Dr. Robert H. L. Jack, T.D., is interested in a wide range of plants, through trees, native and exotic, to shrubs, but particularly in rhododendrons, and especially in species rhododendrons. He lives in the south of Scotland, about 30 miles south-west of Edinburgh, which is a delightful but cold and exposed part of the country. Early autumn frosts and sharp spring frosts do not permit him to grow a wide range of rhododendrons, but closeness to the Royal Botanic Garden in Edinburgh compensates. He has visited many other gardens, and is a committee member of the Rhododendron, Camellia and Magnolia Group of the R.H.S.

Mark Jury is son of Felix Jury, one of New Zealand's foremost plant breeders. With his father, he continues to develop the seven-acre garden on the Taranaki coast of New Zealand, and is now flowering and trialling many of his own rhododendron hybrids. He and his wife own a small plant nursery, supplying unusual material by mail order. Rhododendrons, particularly the *Maddenia* and *Vireya* subsections, are their speciality.

Guan Kaiyun graduated from the Yunnan Teachers' University in 1975. He is a Senior Research Fellow and Assistant Director at the Kunming Institute of Botany at the Academy of Sciences of China. He is also the Vice-President of the Rhododendron Society of China and Director of the Chinese Science and Technology Translators' Association. He works as a taxonomic botanist, in particular studying the families *Dioscoreaceae* and *Smilacaceae*, and is also very interested in the genera *Rhododendron* and *Camellia*. He contributed to *Yunnan Camellias of China* and *Rhododendrons of China* and translated them into English. He has travelled widely in China and the rest of the world.

Mervyn S. Kessel lives in Argyll, Scotland, and works for the District Council's Environmental Services Department. He has been involved in professional horticulture for over 30 years, since his apprenticeship with Glasgow Parks Department. For several years he worked at the Glasgow Botanic Gardens and later studied at the West of Scotland Agricultural College Auchincruive and the Royal Botanic Garden, Edinburgh. He has written a book on rhododendrons and azaleas, published in 1981, as well as numerous articles for various horti-

cultural journals. He is the President of the Scottish Rhododendron Committee, a member of the NTS Advisory Committee for Arduaine Gardens and the Principal External Verifier for Horticultural Qualifications operated by the Scottish Vocational Educational Council.

David Knott studied horticulture at the Royal Botanic Garden, Edinburgh, Scotland, and has been Assistant Curator of Dawyck Botanic Garden near Peebles since 1992. Prior to this he was Garden Supervisor at Logan Botanic Garden near Stranraer. Both gardens are specialist gardens of the Royal Botanic Garden, Edinburgh. He has also worked in other well-known rhododendron gardens, including Brodick Castle Gardens on the Isle of Arran and Castle Kennedy Gardens near Stranraer. During the autumn of 1991 he travelled to western Sichuan where he saw many rhododendron species in the wild.

Isobyl I. La Croix graduated from Edinburgh University, Scotland, where she first became interested in rhododendrons. She worked for two years as Assistant Lecturer in Agriculture at Aberdeen University, then lived in Ghana and Kenya. After their return, she and her husband became active members of the RHS Rhododendron and Camellia Group. In 1973 she wrote the book *Rhododendrons and Azaleas*, as well as numerous articles for *Country Life* and *Amateur Gardening*, among others. In 1978 she wrote *Gardening in the Shade*. She then lived in Malawi for ten years, but during that time went on a Rhododendron Group tour of Cornwall, which she wrote up for the *Yearbook*. In Malawi, her interest gravitated towards orchids and she has since written a book about them, has an Honorary Research Associateship at Kew and they are one of the specialities of the small nursery she now runs with her husband in Gairloch, North-West Scotland, near Inverewe Gardens. Here they have planted over 2000 trees, and are waiting for them to grow high enough to provide the shelter for rhododendrons which she still grows from seed.

Roy Lancaster has been interested in travel and the world's wild places for as long as he can remember. Nine visits to China and expeditions and treks to other countries including Bhutan and Nepal have acquainted him with a wide range of rhododendron species and their associated floras. For 18 years he worked for the well-known nursery firm Hilliers and for ten years was the First Curator of the Hillier Gardens and Arboretum, Hampshire, UK. He has written several books including *Travels in China – A Plantsman's Paradise*, and played a major role in the publication of Hillier's *Manual of Trees and Shrubs*. In 1972, he was awarded the Gold Veitch Memorial Medal for Services to Horticulture by the Royal Horticulture Society and, in 1988, he was awarded the Royal Horticultural Society's most prestigious award, the Victoria Medal of Honour.

David Leach first became enamoured of rhododendrons in 1944, when visiting the collection of Dr. G. O. Clark at Newburyport, Massachusetts, USA. When he discovered that there were many hundreds of species with astounding variations, his education as a geneticist led him to recognize the extraordinary opportunities for hybridizing. He began an intensive study of rhododendrons in the United States and overseas, which eventually resulted in his book *Rhododendrons of the World*. His hybridization programme started in 1946, and is now in its fifth generation. Many hundreds of thousands of seedlings yielded 81 clones, named and distributed as different and superior hybrids to those existing in commerce in very cold climates. He undertook a number of research projects, several of which are still underway. His efforts have been recognized with two honorary doctorates, six gold medals and numerous awards including the Loder Cup from the Royal Horticultural Society. Work for him has been a gratifying combination of science and aesthetics − he is thankful that he was able to make his vocation a full-time occupation.

Beverley Lear runs a botanical and landscape conservation consultancy with her husband. As Lear Associates, they have pioneered the development of plant catalogues and their use in preparation of management plans and interpretive material for visitors. They have worked in gardens and plant collections throughout Britain for the National Trust and private estates, focusing on the study of woody plants in cultivation and in particular the recording and identification of rhododendrons, including the compilation of a complete record of the rhododendrons at Exbury, and work on hardy hybrid rhododendrons in other collections.

Kenneth Lowes was born on the Northumberland Pennines in England, and his earliest recollections include the resinous scent of pinewoods nearby. There was little amenity gardening in the area, but skilful cultivation of very hardy vegetables was essential for survival, and belief in the generous use of farm manure was universal. Exhibiting and competing were natural activities. So when, in August 1939, Gunner Lowes of the Territorial Army was called up, he was away at Southport Show, exhibiting gladioli. After six years of war followed by five years at university, by now married and living in the Home Counties, he began flower gardening again. Not until 1970 was he overcome by rhododendron fever. This looks like being terminal, and no attempt is being made to alleviate the affliction.

During the last 25 years he has amassed some thousands of slides of woodland gardening and rhododendrons, given slide-lectures, edited the *Quarterly Bulletin* of the RHS Rhododendron Group for three years, and written various articles for RHS Yearbooks and other gardening publications.

His wife and two daughters have evaded rhododendron fever, and although they all suffer from less potent horticultural maladies, they continue to support him.

Ronald J. D. McBeath is Assistant Curator at the Royal Botanic Garden, Edinburgh, Scotland. His responsibilities include the maintenance and development of the Rock, Alpine and Peat Gardens where the internationally renowned collection of dwarf rhododendrons is grown. To enhance the collections in the garden and to study rhododendrons in their natural habitats, he has participated in five botanical expeditions to China, one to Sikkim and four to Nepal. He has led eight botanical holiday treks to the Himalayas. He is actively involved in the Scottish Rock Garden Club and Alpine Garden Society. His main interests are bird watching, studying all aspects of Scottish wildlife and hill walking in Scotland.

Edward G. Millais started his horticultural career at Knap Hill Nursery in 1937. After the war, he and his wife Rosemary bought Crosswater Farm in Surrey, England, where for the first 20 years they grew mushrooms, but at the same time started developing their rhododendron garden. In 1967 they began growing rhododendrons commercially, and in 1982−3 they went on expeditions to Sikkim and Nepal with Mrs. Daku Tensing, the wife of the famous Everest climber. In 1987 they visited Canshan and Yulongshan in Yunnan, and in 1988 they joined an expedition with Keith Rushforth to Bhutan. In 1990 Edward Millais organized an expedition to Gonggashan and Kanting (Tatsien-lu) in Sichuan, and in 1992 and 1994 two further expeditions, both to the Salween/Mekong Divide. He is a member of the RHS Rhododendron and Camellia Committee, has written occasional articles for the Society's Yearbook and done a certain amount of lecturing. He has also done some hybridizing of rhododendrons and azaleas, and some of the results have achieved an Award of Merit and gone into production. In 1991 he retired and handed the running of the nursery over to his son David.

Robert J. Mitchell is the Property Manager at Branklyn for the National Trust for Scotland, a garden which features many species of rhododendron, especially those from the Sino-Himalayas. For 25 years he was Curator of the St. Andrews University Botanic Garden and now is Honorary Curator/Consultant. His interest in rhododendrons was nurtured at the Royal Botanic Garden, Edinburgh, where he trained, and was then appointed to the staff in charge of rhododendrons and other woody plants. He was Editor of the *Scottish Rock Garden Club Journal* for seven years, but his crowning glory was as leader of the Sino-British Expedition to Cangshan in 1981 that subsequently opened up South-West China to further expeditions through joint partnerships with the Kunming Institute of Botany.

Sonja Nelson has been editing the *American Rhododendron Society Journal* since 1991. Her background is in newspaper journalism, but she has been an avid gardener for 25 years. When she moved to the North-West of the U.S.A., she became particularly interested in rhododendrons, and her work with the American Rhododendron Society has deepened her interest further. Through her job as journal editor she has visited numerous public and private rhododendron gardens, which have impressed her with the diversity of the species and hybrids being grown today.

Rick Petersen received his Bachelor of Science degree in Botany from the University of Washington, Seattle, Washington, in 1981. He began work at the Rhododendron Species Foundation in Washington, USA, in 1985 and now holds the position of Garden Manager. During his work at the Foundation, species rhododendrons have become a source of intense fascination for him, due to their diversity and beauty.

Nigel Price is Head Gardener at Brodick Castle on the Isle of Arran, Scotland. A keen, all-round plantsman, rhododendrons are his first love, originally inspired by Robert Stephenson Clarke's collection at Borde Hill in West Sussex, England. After nearly seven years at Inverewe Gardens in the far North-West of Scotland, he moved in 1990 to his present position at Brodick, an 80-acre woodland garden with a world-famous rhododendron collection. Owned by the National Trust for Scotland, the garden holds the National Collections of Subsections *Grandia, Falconera* and *Maddenia*. Apart from continuing to maintain and develop the property and its plant collection, he is keen to promote the garden to visitors of all levels of interest through increased access and interpretation. He is an occasional contributor for the R.H.S. Rhododendron, Magnolia and Camellia Society's *Yearbook* and a member of the Scottish Rhododendron Society.

Edmund de Rothschild followed in the footsteps of his father, Lionel de Rothschild, the founder and architect of the Exbury Gardens in Hampshire, England − a most daunting task. After the neglect of the long years of World War II, his mother, with the help of Freddie Wynniat, their late beloved Head Gardener, and assisted by the then Agent, Peter Barber, the co-author of *The Rothschild Rhododendrons*, began the major task of clearing and replanting, using his father's best hybrids. Initially, he was preoccupied with the family bank and other projects. After his mother's death in 1975, he became more actively involved in the Gardens and continued the programme of hybridization. Many of these hybrids have received awards. With the help of Douglas Betteridge, his present and very distinguished Head Gardener, and an excellent small staff, he further developed the Gardens, which successfully weathered the devastation caused by the great storms of 1987 and 1990. More recently, with generous support from his brother Leopold, a charitable trust has been formed so that in future the thousands of visitors that come to the Gardens can continue to enjoy the stupendous show of colour in the spring and the peace and tranquillity of the Gardens all the year round.

Keith Rushforth has a forestry background and practises as an Arboricultural Consultant, which he describes as "anything to do with trees without actually doing anything to them". He has also published books on trees. He has travelled to Bhutan, China, Vietnam (where he saw many wonderful rhododendron species) and Mexico (where unfortunately he didn't find any!). He is interested in all woody plants, especially those that are hardy in the United Kingdom. Among rhododendrons his particular interest is the tree species, especially the *Grandia* and *Falconera* subsections.

Tony Schilling, M.Arb, F.I.Hort, F.L.S., F.R.G.S., V.M.H., joined the staff of the Royal Botanic Gardens, Kew, in 1959 and in 1967, following his secondment to the Royal Nepal Government as an advisor for the Royal Botanic Gardens in the Kathmandu valley, he took over the management of Wakehurst Place ("Kew in the countryside") in West Sussex, England − a post he held until his early retirement in 1991. During his 32 years at Kew he took part in numerous plant-hunting expeditions to Nepal, Bhutan and China, as well as to other parts of the world, including Arctic Norway, Poland, Slovakia, Spain, Portugal, Greece, North America, New Zealand and Australia. He is now in partnership with his wife as a horticultural consultant, as well as writing and lecturing on a wide range of botanical and gardening subjects. He has published over 100 specialist papers and articles for a variety of scientific and learned journals, and contributes regularly to magazines such as *Country Life*. He is a Trustee of the Tree Register of the British Isles and a member of six RHS Committees, including the Rhododendron and Camellia Committee. He is also a member of the American Rhododendron Society (Scottish Chapter). In 1982 he was awarded the Loder Cup (awarded annually by the RHS for achievement in work associated with the genus *Rhododendron*), and in 1990 he was awarded the Victoria Medal of Honour.

Ian W. J. Sinclair became Garden Supervisor with the Royal Botanic Garden, Edinburgh, Scotland, in 1981. He worked in temperate and tropical propagation with the Living Study Collections and in the nursery at Edinburgh before transferring to the Younger Botanic Garden, Benmore, in 1986, where the main collections are conifers and rhododendrons. He has been involved in the development of the Bhutanese Glade. He has travelled to Borneo (Sarawak and Sabah) in 1982, to

Bhutan in 1984, 1990 and 1993 and to China (North-West Yunnan) in 1992. He has written for the *IDS Yearbook*, the *R.H.S. Rhododendron Yearbook* and *The Plantsman*, and co-written *The Benmore Discovery Trail*, as well as contributing to the Zhongdian Prefecture Proposals Towards an Integrated Strategy for Conservation, Tourism and Forestry. He lectures nationally and internationally.

Graham Smith has been Curator and Horticultural Director of the Pukeiti Rhododendron Trust, New Zealand, since 1969. He was educated in England and joined the Royal Parks apprenticeship scheme in its inaugural year, spending five years at Regent's Park before going to Kew for the Diploma course. He came to New Zealand at the end of 1968 and has enjoyed the challenge of managing Pukeiti as a unique rainforest garden ever since. In that time, the rhododendron collection has tripled and special emphasis has been placed on species from wild source collections. Pukeiti's climate allows for a very wide range of rhododendrons to be grown but is particularly suited to the *Grandia*, *Maddenia*, *Arborea* and *Vireya* sections. The latter collection, in a covered bush walk, is one of the finest in the world and is the result of Graham's interest in these special plants. Graham has led tours to many parts of the world and has collected rhododendrons in Papua New Guinea, Malaysia and China. Graham is keen to see an international network of rhododendron gardens specializing in specific sections to ensure the gene pool of resource material is as comprehensive as possible, bearing in mind that wild populations are under threat in many parts of the world.

Prof. Dr. Siegfried Sommer trained as a gardener and landscape gardener, and studied at the Humboldt University of Berlin, Germany. He is now professor at the Technical University of Dresden where he has been teaching on "Plants and their Uses" in the Department of Landscape Architecture since 1959. He was the Chairman of the Working Group on Rhododendrons in the former German Democratic Republic's Cultural League, and has been a member of the German Rhododendron Society since 1990. He co-wrote a book on rhododendron species and hybrids, reflecting his main interests: wild rhododendron species, cultivars and their uses in parks and gardens. In 1988 he travelled to North Korea.

Herbert Spady is a retired orthopaedic surgeon. He is a member or associate member of various chapters of the American Rhododendron Society (ARS) and has held a number of positions for them, and is a member of the Rhododendron Species Foundation (RSF). He was awarded the Gold Medal by the ARS in 1991. He is owner and keeper of a four-acre rhododendron garden and a keen hybridizer – he registered 'Honsu's Baby'. He has contributed many articles to the *Rhododendron News* and

the ARS *Journal (Bulletin)*, as well as publishing the RSF's *Rhododendron Species Dictionary* in 1985. His lecture programmes include the RSF China Tour of 1988–9 and Trekking Sikkim 1992.

Prof. Dr. Wolfgang Spethmann is Professor for Nursery Science at the University of Hanover, Germany, and Acting Director of the Institute for Fruit and Nursery Science. He specializes in propagation by cuttings of difficult genera and species (oak, rhododendron, roses, bamboo), the cultivation of rhododendrons with peat, N-fertilization and the comparison of propagation methods of rhododendrons (cuttings, graftings, *in-vitro* plants). He organized the 5th International Rhododendron Conference in Bad Zwischenahn in 1992 and edited the proceedings. He is also the author of the most comprehensive international *Rhododendron Bibliography*. He is a member of the steering committee of the International Rhododendron Union and, since 1984, he has been a member of the board of directors of the German Rhododendron Society and the editor of their journal, *Immergrüne Blatter*.

Major Thomas Le M. Spring-Smyth was a regular Royal Engineers Officer serving with the Indian Army in India and Burma and later with the Gurkha Engineers in Malaya and Hong Kong. Always a keen gardener, a trek in West Nepal in 1955 revealed the richness and variety of Himalayan plants to him – and he was determined to return. After retirement in 1960 he joined a scientific collecting expedition from the British Museum (Natural History) as an unpaid freelance collector and interpreter in 1961–2. He concentrated on the seed of rhododendrons and other high-altitude plants. His mother, a knowledgeable plantswoman, distributed seed to Kew to the Royal Botanic Garden, Edinburgh, to the Savill Gardens, and sowed some herself. While working for the United Nations and British Government in Nepal from 1962 to 1973, he had further scope for collecting seed and a variety of live plants. He is currently a member of the RHS Rhododendron and Camellia Committee and the Rhododendron, Camellia and Magnolia Group Committee.

Ivor T. Stokes is Curator for the three sites that make up the City of Swansea's Botanical Complex, Wales. He is responsible for Plantasia, a new high-tech tropical glasshouse in the city centre, and the long-established Botanic Gardens in Singleton Park. He has been leading the restoration of the fine rhododendron collection at Clyne Castle, Swansea, which was originally planted by Admiral A. Walker-Heneage-Vivian during the first half of this century, but had fallen into obscurity after the Admiral's death in 1952. He has put Clyne back on the map of famous rhododendron gardens by showing many unusual and spectacular species as well as the hybrids bred at Clyne at RHS shows in London. Clyne holds,

amongst others, the NCCPG collections of rhododendron Subsections *Triflora* and *Falconera*. In 1993 he travelled to Yunnan in western China where he saw and collected rhododendrons in the wild.

Dr. Noel Sullivan describes his interest in rhododendrons as an obsession, with an equal infatuation for species and hybrids of distinction. Formerly a private gardener, he now enjoys retirement as Curator of the Emu Valley Rhododendron Garden in Burnie, Tasmania – one of the world's mildest climates suitable for growing rhododendrons. Here he is able to follow his interest in and appreciation of the rhododendron species and grows most with success. He has collected rhododendrons in Sikkim, and registered a form of *R. arboreum* as ssp. *arboreum* f. Stowport with the International Rhododendron Register.

Hideo Suzuki has been Vice-President of the Japanese Rhododendron Society for almost 20 years. He is also the Director of the International Rhododendron Union and Director and Founding Member of the Japan Branch of the Royal Horticultural Society. In 1982, the Gold Medal Award was conferred on him by the American Rhododendron Society in Washington, USA, and in 1994 he received the Veitch Memorial Gold Medal Award from the Royal Horticultural Society, London, UK. He is keenly interested in Japanese native azaleas and often speaks and writes about them and other subjects, both in Japan and overseas.

Margaret Tapley created, over 20 years, a woodland garden, specializing in rhododendrons and camellias, at Pigeon Bay on Banks Peninsula, New Zealand, which has been used extensively for visiting and community activities and is still open to the public. She now works as a Landscape Garden Consultant and for nurseries in Christchurch, New Zealand, and is a member of the New Zealand Camellia Society, the New Zealand Rhododendron Society and the Royal Horticultural Society. She has travelled widely overseas, visited many gardens, and gives talks on gardening topics throughout New Zealand. She has written three books on rhododendrons and camellias.

Michael & Sue Thornley are architects, working in Glasgow, Scotland. They bought Glenarn in Rhu, 30 miles west of Glasgow, and spent the last ten years clearing and restoring it. Glenarn was laid out in the 1850s, but the wonderful rhododendron collection was built up by Archie and Sandy Gibson from 1927 to 82. In addition, it is a fine woodland garden with magnolias, embothriums, primulas and bulbs in season, open to the public under Scotland's Gardens Scheme, daily from 21 March to 21 May each year. They share a keen interest in gardens and plants, and have only become deeply involved with rhododendrons in the last ten years. Michael was the editor of the

American Rhododendron Society Newsletter, Scottish Chapter, for the first five years of its existence and, with Sue, has spoken to horticultural and other societies on Glenarn and rhododendrons.

Scott Vergara is Executive Director of the Rhododendron Species Foundation in Washington, USA. Interested in plants from the age of six, when he helped a cousin transplant bedding plants in a greenhouse, he has never got "the dirt out from under my fingernails". While studying ornamental plant breeding and genetics at college, he became interested in rhododendrons. Realizing that the thousands of hybrids in existence then only used a small portion of the tremendous genetic potential of the genus, he began to investigate the vast range of plant characteristics (such as habit, leaf size, texture, flower colour, size, fragrance, pest resistance and environmental tolerance) that were embodied in the 850-plus species. He is particularly curious about the basic genetics of the plants with which he works and enjoys seeking superior forms of species to provide plant breeders with the genetic material to create even more incredible hybrids. While he appreciates the opulence of many hybrids, he is most attracted to the simplicity of form and the degree of adaptation that the ancient species rhododendrons exemplify – his link to a time long forgotten.

Maurice Wilkins became interested in rhododendrons a long while ago when he was working in a woodland garden in the south of England. Later, after a few years of rhododendron-less gardening in the east of Scotland, he took up the post of Head Gardener at Ross Priory, owned by the University of Strathclyde, a garden with an interesting collection of both species and hybrids, including many of the old hardy hybrids. In 1992 he joined the National Trust for Scotland as Head Gardener at Arduaine in Argyll. This coastal garden, established at the turn of the century by J. A. Campbell, was recently gifted to the Trust by the Wright brothers, and has a fine species collection.

F. J. Williams, C.B.E., D.L., has lived at Caerhays Castle in Cornwall, England, since 1955. This garden was created by John Charles Williams (1860–1939), who was one of the first people to test out the plants brought back from China by E. H. Wilson. The garden later became the recipient of seeds sent back to the UK by George Forrest. Despite the tribulations of the weather, many of the original plants still survive. The aim of Julian Williams, and his son Charles, is to maintain the garden as a centre for rare and beautiful plants.

PARKS AND GARDENS

AUSTRALIA
(selected by Lesley Eaton)
Emu Valley Rhododendron Gardens, Burnie, Tasmania 7320; Tel. +61 (0)3-751 1980

Illawarra Rhododendron Park, Mt. Pleasant, Wollongong, N.S.W., +61 (0)4-330 0027

Mt. Lofty Botanic Gardens, Lampert Road, Piccadilly, S.A.; +61 (0)8 -339 2707

Olinda Rhododendron Garden, Georgian Road, Olinda; Tel. +61 (0)3-751 1980

Royal Botanical Gardens Melbourne, Birdwood Avenue, South Yarra, Victoria 3141; Tel.: +61 (0)3-655 2300

Royal Botanical Gardens Sydney, Mrs. Macquarie Road, Sydney, N.S. W. 2000; Tel. +61 (0)2-231 811

Royal Tasmanian Botanical Gardens, Queens Domain, Hobart, Tasmania 7000; Tel. +61 (0)2-346 299

Australia also has an Open Garden Scheme, with a guide book published each year. Lesley Eaton's own garden Kalbar/Shangri La is listed in this book.

CANADA
(selected by Sonja Nelson)
University of British Columbia Botanical Garden, 6501 NW Marine Drive, Vancouver, B.C. V6T IW5; Tel. +1-604-228-4186

Van Dusen Botanical Garden, 5251 Oak St., Vancouver, B.C. V6M 4H1, Canada; Tel. +1-604-266-7194

SWEDEN
(selected by Dr. Björn Aldén)
The Gardens of Sofiero, Helsingborg, South Sweden (formally the collection of the late King Gustav VI Adolf); Tel. +46 (0)42-13 74 00

Göteborg Botanical Garden, Carl Skottsbergs gata 22, Göteborg, Sweden; Tel. +46 (0)31-41 37 50

GERMANY
(selected by Prof. Dr. W. Spethmann)
Botanischer Garten und Rhododendron-Park, Marcusallee 60, 28359 Bremen, Germany; Tel. +49 (0)421-361 3025

Lehr- und Versuchsanstalt für Gartenbau, Rostrup, 26160 Bad Zwischenahn, Germany; Tel. +49 (0) 4403-97960

Rhododendron-Waldpark und Baumschule Hobbie, Linswege-Petersfeld, 26665 Westerstede, Germany; Tel. +49 (0)44 58-201

Rhododendron-Park Gristede der Baumschule Bruhns, 26160 bad Zwischenahn, Germany; Tel. +49 (0) 4403-6010

GREAT BRITAIN
(selected by Bruce Archibold and Marianna Kneller)
Home Counties and South-East:
Borde Hill Garden, Haywards Heath, West Sussex; Tel. +44 (0)1444-450 26

Exbury Gardens and Nurseries, Exbury nr. Southampton, Hampshire; Tel. +44 (0)1703-891203

The Great Park, Windsor, Berkshire; Tel. +44 (0)1753-860222

Leonardslee Gardens, Lower Beeding, Horsham, West Sussex; Tel. +44 (0)140 376-212
Nymans, Handcross, West Sussex; Tel. +44 (0)1444-400321
Wakehurst Place, Ardingly, West Sussex; Tel. +44 (0)1444-892701

South-West:
Caerhays Castle, Gorran, St. Austell, Cornwall; Tel.: +44 (0)1 872 501250
Lanhydrock nr. Bodmin, Cornwall: Tel. +44 (0)1208-3320

Minterne, Cerne Abbas, Dorset; Tel. +44 (0)13003-370

Tremeer Gardens, St. Tudy, Bodmin, Cornwall; Tel. +44 (0)11208-850313

Trengwainton Gardens, nr. Penzance, Cornwall; Tel. +44 (0)1736-63148

Trewithen, Probus, Cornwall; Tel. +44 (0)1726-882418/882585
Woodland Grove, Bovey Tracey, Devon

Midlands and Lake District:
The Dorothy Clive Garden, Willoughbridge, Staffordshire; Tel. +44 (0)1782-680322

Lea Rhododendron Gardens, Lea, Matlock, Derbyshire; Tel. +44 (0)162 984-380

Lingholme, Keswick, Cumbria; Tel. +44 (0)1596-72 003

Muncaster Castle, Ravenglass, Cumbria; Tel. +44 (0)16577-614/203

Ness Gardens, Cheshire; Tel. +44 (0)151-336 2135

Wales:
Bodnant Garden, Tal-y-Cafn, nr. Colwyn Bay, Clwyd, Wales; Tel.: +44 (0)1492-650460

Clyne Castle Gardens, Swansea, Wales
Portmeirion, Penrhyndenraeth, Gwynedd, Wales; Tel. +44 (0)1766-770228
Powis Castle Gardens, Welshpool, Powys; Tel. +44 (0)1938-554336

Scotland:
Arduaine Gardens, Kilmelford, by Oban, Argyll PA34 4XQ; Tel. +44 (0)18522-00366

Blackhills, by Elgin, Morayshire; Tel. +44 (0)1343-842223

Brodick Castle, Brodick, Isle of Arran, KA27 8HY; Tel. +44 (0)1770-302 202, Fax +44 (0)1770-302312

Castle Kennedy Gardens, Castle Kennedy, Stranrae, Wigtownshire DG9 8BX; Tel. +44 (0)1776-2024

Crarae Glen Garden, Minard, Strathclyde; Tel. +44 (0)1546-86633

Corsock House, Castle Douglas, Dumfries
Dawyck Botanic Garden, Stobo, Peebles, Borders; Tel. +44 (0)1721-6254

Glenarn, Rhu, Dunbartonshire G84 8LL; Tel. +44 (0)1436-820493

Inverewe Garden, Poolewe, Achnasheen, Ross-shire IV22 2LG; Tel. +44 (0)144 586-200

Logan Botanic Garden, Port Logan, Wigtownshire DG9 9ND; Tel. +44 (0)1776 86-231

Royal Botanic Garden, 20A Inverleith Row, Edinburgh EH3 5LR; Tel. +44 (0)131-552 7171

Stonefield Castle, Tarbert, Knapdale; Tel. +44 (0)1880-820836

The Younger Botanic Garden, Benmore by Dunoon, Argyll PA23 8QU, Scotland; Tel. +44 (0)01369-6261

IRELAND AND NORTHERN IRELAND
Glenveagh National Park, Churchill, Letterkenny, Co. Donegal, Ireland; Tel. +353-(0)74-37090 and (0)74-37262

John F. Kennedy Park, New Ross, Co. Wexford, Ireland; Tel. +353-(0)51-88171

Malahide Castle, Malahide, Co. Dublin, Ireland; Tel. +353-(0)1-846 2516
Mount Congreve, Kilmeadon, Co. Waterford, Ireland; Tel. +353-(0)51-384115

Mount Stewart, Greyabbey, Newtownards, Co. Down, Northern Ireland; Tel. +44-(0) 12477-88387

Mount Usher, Ashford, Co. Wicklow, Ireland; Tel. +353-(0)404-40116

NEW ZEALAND
(selected by Graham Smith)
Crosshills Gardens, Kimbolton, Manawatu; Tel. +64-6-328-5797

Dunedin Botanic Garden, P.O. Box 5045, Dunedin, Otago; Tel. +64-3-477 4000

Mark Jury Nursery, Tikorangi, RD 43, Waitara, North Taranaki, New Zealand; Tel. +64 (0)6-7548 577

Kimbolton Rhododendron Park, Kimbolton, Manawatu

Koromiko Nursery, RD 9, Whangarei, New Zealand

Pukeiti Rhododendron Trust, RD 4, New Plymouth, Taranaki; Tel./fax +64-67-524 141

Tannock Glen Garden, Dunedin; Tel. +64-3-477 4000

USA
(compiled by Sonja Nelson)
Arnold Arboretum, 125 Arborway, Jamaica Plain, MA 02130-3519; Tel. +1-617-524-1718

Bartlett Arboretum, University of Connecticut, 151 Brookdale Road, Stamford, CT 06903; Tel. +1-203-322-6971

Bloedel Reserve, 7571 NE Dolphin Drive, Bainbridge, WA 98110; Tel. +1-206-842-7631

Brooklyn Botanic Garden, 1000 Washington Ave, Brooklyn, NY 11225; Tel. +1-718-622-4433

Callaway Gardens, U.S. Highway 27, Pine Mountain, GA 31822; Tel. +1-404-663-5154

Cecil & Molly Smith Garden, 5065 Ray Bell Road, St. Paul, OR 97137; Tel. +1-503-246-3710

Crystal Spring Garden, Portland, OR

Dallas Arboretum, 8617 Garland Road, Dallas, TX 752; Tel. +1-214-327-8263

Holden Arboretum, 9500 Sperry Road, Mentor, OH 44060; Tel. +1-216-946-4400

Jenkins Arboretum, 631 Berwyn Baptist Road, Devon, PA 19333; Tel. +1-215-647-8870

Meerkerk Rhododendron Gardens, PO Box 154, Greenbank, WA 98253; Tel. +1-206-678-8740

Mendocino Coast Botanical Gardens, PO Box 1143, Fort Bragg, CA 95437; Tel. +1-707-964-4352

Morris Arboretum, 9414 Meadowbrook Road, Philadelphia, PA 19118; Tel. +1-215-247-5777

National Arboretum, 3501 New York Ave., NE, Washington, DC 20002; Tel. +1-202-457-4831

New York Botanical Garden, Bronx, NY 10458; Tel. +1-212-220-8700

Planting Fields Arboretum, Planting Fields Road, Oyster Bay, Long Island, NY 11771; Tel. +1-516-922-9206

Rhododendron Species Botanical Garden, PO Box 3798 Federal Way, WA; Tel. +1-206-838-4646

Strybing Arboretum and Botanical Gardens, 9th Ave., and Lincoln Way, San Francisco, CA 94122; Tel. +1-415-661-1316

University of California, Berkeley Botanical Garden, Centennial Drive, Berkeley, CA 94720; Tel. +1-415-643-8040

University of California-Los Angeles, UCLA, Los Angeles, CA 90024; Tel. +1-213-825-3620

University of Maine, Orono, Maine +1-04469-5722

University of North Carolina at Charlotte Botanical Gardens, Charlotte, NC 28223; Tel. +1-704-547-2555

Washington Park Arboretum, University of Washington XD-10, Seattle, WA 98195; Tel. +1-206-325-4510

Western North Carolina Arboretum, Route 3, Box 1249B, Asheville, NC 28806. Tel. +1-704-665-2492

Winterthur Museum, Inc. Winterthur, DE 19735; Tel. +1-302-656-8591

RHODODENDRON SOCIETIES AND ASSOCIATIONS

Australia
The Australian Rhododendron Society, The Secretary, P.O. Box 21, Olinda, Victoria 3788. Yearly journal.

Canada
Rhododendron Society of Canada, 4271 Lakeshore Road, Burlington, Ontario

Germany
Deutsche Rhododendron-Gesellschaft e.V., Marcusallee 60, 28359 Bremen 17, Germany; Tel. +49 (0)421-361 3025; annual publications *Immergrüne Blätter, Rhododendron Jahrbuch*.

Great Britain
The Rhododendron and Camellia Group of the Royal Horticultural Society, Mrs Joey Warren, Netherton, Buckland Monachorum, Yelverton, PL20 7NL; yearbook: *Rhododendrons with Magnolias and Camellias*

Japan
The Japanese Rhododendron Society, Teruo Takeuchi, 8-5, 2-chome Goshozuka, Takatsu-k, Kawasaki, Japan

New Zealand
The New Zealand Rhododendron Association, P.O. Box 28 Palmerston North, New Zealand

Dunedin Rhododendron Group, S. J. Grant, 25 Pollack St., Dunedin, New Zealand

Pukeiti Rhododendron Trust, Carrington Rd., New Plymouth, New Zealand

USA
The American Rhododendron Society; publishes *Journal*

The Rhododendron Species Foundation, P.O. Box 3798, Federal Way, WA 98063-3798, USA; publishes *Rhododendron Notes and Records*

Left *R. glaucophyllum* Rehd. "with a bluish grey leaf"; Right

R. charitopes Balf. f & Farr. "with a graceful appearance", ssp.

Tsangpoense

SPECIES LIST

PART I SUBGENUS HYMENANTHES

The plants in this subgenus are the elepidote (without scales) rhododendrons. Their leaves are evergreen, and, along with the young shoots, can be covered with a waxy bloom (glaucous), be free of hairs (glabrous) or have varying degrees of a hairy covering on the under-surface of the leaves (indumentum). They are only rarely aromatic. The buds are usually revolute and the calyx varies from obsolete to large and beautifully formed (e.g. *R. thomsonii*). The inflorescence is terminal raceme. Corollas are 5–10 lobed, saucer-shaped to tubular-campanulate, and can be with or without nectar pouches. The stamens have between 10 and 20 chambers. The seeds may be with or without wings. The capsules have hard woody valves.

(Subgenus *Hymenanthes* has one section, *Pontica*, which has approximately 220 species divided into 24 subsections by Dr. D. F. Chamberlain in his *Revision of Rhododendron II, Subgenus Hymenanthes vol. 39. No 2.* (1982).)

SUBSECTION **ARBOREA** Sleumer
Syn: *Arboreum*, Subseries *Arboreum*

R. arboreum
  ssp. *arboreum*
  ssp. *cinnamomeum* (Wall. ex Lindl.) Tagg
    var. *cinnamomeum* (var. album)
    var. *roseum* Lindl.
  ssp. *delavayi*
    var. *delavayi* (Franch.) Chambln.
    var. *peramoenum* (Balf. f. & Forr.) Chambln.
  ssp. *nilagiricum* (Zenker.) Tagg
  ssp. *zeylanicum* (Booth.) Tagg
R. lanigerum Tagg
R. niveum Hooker f.

SUBSECTION **ARGYROPHYLLA** Sleumer
Syn: Series *Arboreum*, Subseries *Argyrophyllum*

R. adenopodum Franch.
R. argyrophyllum Franch.
  ssp. *argyrophyllum*
  ssp. *hypoglaucum*(Hemsl.) Chambln.
  ssp. *omeiense* (Rehd. & Wils.) Chambln.
  ssp. *nankingense* (Cowan) Chambln.
R. coryanum Tagg & Forr.
R. denudatum Lévl.
R. farinosum Lévl. Q.
R. floribundum Franch.
R. formosanum Hemsl.
R. haofui Chun & Fang Q.
R. hunnewellianum Rehd. & Wils.
  ssp. *hunnewellianum* Rehd. & Wils.
  ssp. *rockii* (Wils.) Chambln.
R. insigne Hemsl. & Wils.
R. longipes Rehd. & Wils. Q.
  var. *longipes* Rehd. & Wils.
  var. *chienianum* (Fang Q.) Chambln.
R. pingianum Fang
R. ririei Hemsl. & Wils.
R. simiarum Hance
R. thayerianum Rehd. & Wils.

SUBSECTION **AURICULATA** Sleumer
Syn: *Auriculatum*

R. auriculatum Hemsl.
R. chihsinianum Chun & Fang Q.

SUBSECTION **BARBATA** Sleumer
Syn: Series *Barbatum*, Subseries *Barbatum*

R. barbatum Wall. ex G. Don
R. erosum Cowan
R. exasperatum Tagg
R. smithii Nutt. ex Hooker f.
R. succothii Davidian

SUBSECTION **CAMPANULATA** Sleumer
Syn: Series *Campanulatum*

R. campanulatum D. Don.
  ssp. *campanulatum*
  ssp. *aeruginosum* (Hooker f.) Chambln.
R. wallichii Hooker f.

SUBSECTION **CAMPYLOCARPA** Sleumer
Syn: Series *Thomsonii*, Subseries *Campylocarpum* & Subseries *Souliei*

R. callimorphum
  var. *callimorphum* Balf. f. & W. W. Sm.
  var. *myiagrum* (Balf. f. & Forr.) Chambln.
R. campylocarpum Hooker f.
  ssp. *campylocarpum*
  ssp. *caloxanthum* (Balf. f. & Forr.) Chambln.
R. souliei Franch.
R. wardii W. W. Sm.
  var. *wardii*
  var. *puralbum* (Balf. f. & W. W. Sm.) Chambln.

SUBSECTION **FALCONERA** Sleumer
Syn: Series *Falconeri*

R. basilicum Balf. f. & W. W. Sm.
R. coriaceum Franch.
R. falconeri Hooker f.
  ssp. *falconeri*
  ssp. *eximium* (Nutt.) Chambln.
R. galactinum Balf. f. ex Tagg
R. hodgsonii Hooker f.
R. preptum Balf. f. & Forr.
R. rex Lévl.
  ssp. *rex*
  ssp. *fictolacteum* (Balf. f.) Chambln.
  ssp. *arizelum* (Balf. f. & Forr.) Chambln.
R. rothschildii Davidian

R. semnoides Tagg & Forr.
R. sinofalconeri Balf. f. Q.

SUBSECTION **FORTUNEA** Sleumer
Syn: Series *Fortunei*

R. asterochnoum Diels, Fedd. Rep.
R. calophytum Franch.
  var. *calophytum*
  var. *openshawianum* (Rehd. & Wils. Q.) Chambln.
R. davidii Franch. Q.
R. decorum Franch.
R. diaprepes Balf. f. & W. W. Sm.
R. faithae Chun Q.
R. fortunei Lindl.
  ssp. *fortunei*
  ssp. *discolor* (Franch.) Chambln.
R. glanduliferum Franch. Q.
R. griffithianum Wight
R. hemsleyanum Wils.
R. huianum Fang Q.
R. orbiculare
  ssp. *orbiculare* Decne.
  ssp. *cardiobasis* (Sleumer) Chambln.
R. oreodoxa Franch.
  var. *oreodoxa*
  var. *fargesii* (Franch.) Chambln.
  var. *shensiense* Chambln.
R. platypodum Diels
R. praeteritum Hutch.
R. praevernum Hutch.
R. sutchuenense Franch.
R. vernicosum Franch.

SUBSECTION **FULGENSIA** Chamberlain
Syn: Series *Campanulatum*

R. fulgens Hooker f.
R. miniatum Cowan Q.
R. sherriffii Cowan

SUBSECTION **FULVA** Sleumer
Syn: Series *Fulvum*

R. fulvum Balf. f. & W. W. Sm.
R. uvarifolium Diels

SUBSECTION **GLISCHRA** (Tagg) Chamberlain
Syn: Series *Barbatum*, Subseries *Glischrum*

R. adenosum Davidian
R. crinigerum Franch.
  var. *crinigerum*
  var. *euadenium* Tagg & Forr.
R. glischrum Balf. f. & W. W. Sm.
  ssp. *glischrum*
  ssp. *rude* (Tagg & Forr.) Chambln.
  ssp. *glischroides* (Tagg & Forr.) Chambln.
R. habrotrichum Balf. f. & W. W. Sm.
R. recurvoides Tagg & Kingd.-Wd.
R. vesiculiferum Tagg

SUBSECTION **GRANDIA** Sleumer
Syn: Series *Grande*

R. grande Wight
R. macabeanum Watt ex Balf. f.
R. magnificum Ward
R. montroseanum Davidian
R. praestans Balf. f. & W. W. Sm.

R. protistum Balf. f. & Forr.
  var. *protistum*
  var. *giganteum* (Forr. ex Tagg) Chambln.
R. pudorosum Cowan
R. sidereum Balf. f.
R. sinogrande Balf. f. & W. W. Sm.
R. watsonii Hemsl. & Wils.
R. wattii Cowan

Subsection **GRIERSONIANA** (Davidian ex) Chamberlain
Syn: Series *Griersonianum*

R. griersonianum Balf. f. & Forr.

SUBSECTION **IRRORATA** Sleumer
Syn: Series *Irroratum*

R. aberconwayi Cowan
R. agastum Balf. f. & W. W. Sm.
R. annae Franch.
R. anthosphaerum Diels
R. araiophyllum Balf. f. & W. W. Sm.
R. brevinerve Chun & Fang Q.
R. excelsum Chev. Q.
R. irroratum Franch.
  ssp. *irroratum*
  ssp. *pogonostylum* (Balf. f. & W. W. Sm.) Chambln.
  ssp. *kontumense* (Sleumer Q.) Chambln.
R. kendrickii Nutt.
R. korthalsii Miq. Q.
R. leptopeplum Balf. f. & Forr. Q.
R. lukiangense Franch.
R. mengtszense Balf. f. & W. W. Sm. Q.
R. papillatum Balf. f. & Cooper. Q.
R. ramsdenianum Cowan
R. spanotrichum Balf. f. & W. W. Sm. Q.
R. tanastylum Balf. f. & Ward.
  var. *tanastylum*
  var. *pennivenium* (Balf. f. & Forr. Q.) Chambln.
R. wrayi King & Gamble

SUBSECTION **LANATA** Chamberlain
Syn: Series *Campanulatum*

R. circinatum Cowan & Ward
R. lanatoides Chambln. (provisional name)
R. lanatum Hooker f.
R. tsariense Cowan

SUBSECTION **MACULIFERA** Sleumer
Syn: Series *Barbatum*, Subseries *Maculiferum*

R. longesquamatum C. K. Schneid.
R. maculiferum Franch.
  ssp. *maculiferum*
  ssp. *anhweiense* (Wils.) Chambln.
R. morii Hayata
R. ochraceum Rehd. & Wils. Q.
R. pachysanthum Hayata
R. pachytrichum Franch.
R. pseudochrysanthum Hayata
R. sikangense Fang
R. strigillosum Franch.

SUBSECTION **NERIIFLORA**
Sleumer
Syn: Series *Neriiflorum*

R. *albertsenianum* Forr.
R. *aperantum* Balf. & Ward
R. *beanianum* Cowan
R. *catacosmum* Balf. F. ex Tagg
R. *chamaethomsonii* (Tagg & Forr.)
  Cowan & Davidian
    var. *chamaethomsonii*
    var. *chamaedoron* (Tagg & Forr.)
      Chambln.
    var. *chamaethauma* (Tagg) Cowan
      & Davidian
R. *chionanthum* Tagg & Forr.
R. *citriniflorum* Balf. f. & Forr.
    var. *citriniflorum*
    var. *horaeum* (Balf. f. & Forr.)
      Cowan
R. *coelicum* Balf. f. & Farr.
R. *dichroanthum* Diels
  ssp. *dichroanthum*
  ssp. *apodectum* (Balf. f. & W. W.
    Sm.) Cowan
  ssp. *scyphocalyx* (Balf. f. & Forr.)
    Cowan
  ssp. *septentrionale* Cowan
R. *erastum* Balf. f. & Forr.
R. *euchroum* Balf. f. & Ward
R. *eudoxum* Balf. f. & Forr.
    var. *eudoxum*
    var. *brunneifolium* (Balf. f. & Forr.)
      Chambln.
    var. *mesopolium* (Balf. f. & Forr.)
      Chambln.
R. *floccigerum* Franch.
R. *forrestii* Balf. f. ex Diels.
  ssp. *forrestii*
  ssp. *papillatum* Chambln.
R. *haematodes* Franch.
  ssp. *haematodes*
  ssp. *chaetomallum* (Balf. f. & Forr.)
    Chambln.
R. *mallotum* Balf. f. & Ward
R. *microgynum* Balf. *microgynum*
  Balf. f. & Forr.
R. *neriiflorum* Franch.
  ssp. *neriiflorum*
  ssp. *agetum* (Balf. f. & Forr.) Tagg
  ssp. *phaedropum* (Balf. f. & Forr.)
    Tagg
R. *parmulatum* Cowan
R. *piercei* Davidian
R. *pocophorum* Balf. f. ex Tagg
    var. *pocophorum*
    var. *hemidartum* (Balf. f. ex Tagg)
      Chambln.
R. *sanguineum* Franch.
  ssp. *sanguineum*
    var. *sanguineum*
    var. *haemaleum* (Balf. f. & Forr.)
      Chambln.
    var. *himertum* (Balf. f. & Forr.)
      Chambln.
    var. *cloiophorum* (Balf. f. & Forr.)
      Chambln.
    var. *didymoides* (Tagg & Forr.)
      Chambln.
  ssp. *didymum* (Balf. f. & Forr.)
    Cowan
R. *sperabile* Balf. f. & Forr.
    var. *sperabile*
    var. *weihsiense* Tagg & Forr.
R. *sperabiloides* Tagg & Forr.
R. *trilectorum* Cowan
R. *temenium* Balf. f. & Forr.
    var. *temenium*
    var. *gilvum* (Cowan) Chambln.
    var. *dealbatum* (Cowan) Chambln.

SUBSECTION **PARISHIA** Sleumer
Syn: Series *Irroratum*, Subseries
Parishii

R. *elliottii* Watt ex Brandis
R. *facetum* Balf. f. & Ward
R. *huidongense* R. L. Ming
R. *kyawi* Lace & W. W. Sm.
R. *parishii* C. B. Clarke
R. *schistocalyx* Balf. f. & Forr. Q.

SUBSECTION **PONTICA** Sleumer
Syn: Series *Ponticum*

R. *aureum* Georgi
    var. *aureum*
    var. *hypopitys* (Pojark) Chambln.
R. *brachycarpum* D. Don ex G. Don
  ssp. *brachycarpum*
  ssp. *fauriei* (Franch.) Chambln.
R. *catawbiense* Michx.
R. *caucasicum* Pallas
R. *hyperythrum* Hayata
R. *japonicum* (Blume) C. K.
  Schneid.
    var. *japonicum*
    var. *pentamerum* (Maxim.) Hutch
R. *macrophyllum* D. Don ex G. Don
R. *maximum* L.
R. *ponticum* L.
R. *smirnowii* Trautv.
R. *ungernii* Trautv.
R. *yakushimanum*
  ssp. *yakushimanum* Nakai
  ssp. *makinoi* (Tagg) Chambln.

SUBSECTION **SELENSIA** Sleumer
Syn: Series *Thomsonii*, Subseries
Selense

R. *bainbridgeanum* Tagg & Forr.
R. *calvescens* Balf. f. & Forr.
    var. *calvescens*
    var. *duseimatum*
R. *dasycladoides* Hand.-Mazz.
R. *esetulosum* Balf. f. & Forr.
R. *hirtipes* Tagg
R. *martinianum* Balf. f. & Forr.
R. *selense* Franch.
  ssp. *selense*
  ssp. *dasycladum* (Balf. f. & W. W.
    Sm.) Chambln.
  ssp. *setiferum* (Balf. f. & Forr).
    Chambln.
  ssp. *jucundum* (Balf. f. & W. W.
    Sm.) Chambln.

SUBSECTION **TALIENSIA** Sleumer
Syn: Series *Taliense*, Series *Lacteum*

R. *adenogynum* Diels
R. *aganniphum* Balf. f. & Ward
    var. *aganniphum*
    var. *flavorufum* (Balf. f. & Forr.)
      Chambln.
R. *alutaceum* Balf. F. & W. W. Sm.
    var. *alutaceum*
    var. *iodes*
    var. *russotinctum* (Balf. f. & Forr.)
      Chambln.
R. *balfourianum* Diels
R. *barkamense* Chambln. sp. nov.
R. *bathyphyllum* Balf. f. & Forr.
R. *beesianum* Diels
R. *bureavii* Franch.
R. *clementinae* Forr.
R. *codonanthum* Balf. f. & Forr.
R. *coeloneuron* Diels.
R. *comisteum* Balf. f.
R. *detersile* Franch.
R. *dignabile* Cowan

R. *dumicola* Tagg & Forr.
R. *elegantulum* Tagg & Forr.
R. *faberi* Hemsl.
  ssp. *faberi*
  ssp. *prattii* (Franch.) Chambln.
R. *lacteum* Franch.
R. *mimetes* Tagg & Forr.
R. *nakotiltum* Balf. f. & Forr.
R. *nigroglandulosum* Nitzelius.
R. *phaeochrysum* Balf. f. & W. W.
  Sm.
    var. *phaeochrysum*
    var. *agglutinatum* (Balf. f. & Forr.)
      Chambln.
    var. *levistratum* (Balf. f. & Forr.)
      Chambln.
R. *pomense* Cowan & Davidian
R. *principis* Bur. & Franch.
R. *pronum* Tagg & Forr.
R. *proteoides* Balf. f. & W. W. Sm.
R. *przewalskii* Maxim.
R. *pubicostatum* T. L. Ming.
R. *roxieanum* Forr.
    var. *roxieanum*
    var. *cucullatum* (Hand.-Mazz.)
R. *roxieoides* Chambln. sp. nov.
R. *rufum* Batal.
R. *simulans* (Tagg & Forr.)
  Chambln. (provisional name)
R. *sphaeroblastum* Balf. f. & Forr.
R. *taliense* Franch.
R. *traillianum* Forr. & W. W. Sm.
    var. *traillianum*
    var. *dictyotum* (Balf. f. ex Tagg)
      Chambln.
R. *wasonii* Hemsl. & Wils.
R. *wightii* Hooker f.
R. *wiltonii* Hemsl. & Wils.

SUBSECTION **THOMSONIA**
Sleumer
Syn: Series *Thomsonii*

R. *bonvalotti* Bur. & Franch.
R. *cerasinum* Tagg
R. *cyanocarpum* (Franch.) W. W. Sm.
R. *eclecteum* Balf. f. & Forr.
    var. *eclecteum*
    var. *bellatulum* Balf. f. ex Tagg
R. *eurysiphon*
R. *faucium* Chambln.
R. *hookeri* Nutt.
R. *hylaeum* Balf. f. & Forr.
R. *meddianum* Forr.
    var. *meddianum*
    var. *atrokermesinum* Tagg

R. *populare* Cowan
R. *stewartianum* Diels
R. *subansiriense* Chambln. & Cox
R. *thomsonii* Hooker f.
  ssp. *thomsonii*
  ssp. *lopsangianum* (Cowan)
    Chambln. (provisional name)
R. *viscidifolium* Davidian

SUBSECTION **VENATORA**
Chamberlain
Syn: Series *Irroratum*, Subseries
Parishii

R. *venator* Tagg

SUBSECTION **WILLIAMSIANA**
Chamberlain
Series *Thomsonii*, Subseries
Williamsianum

R. *williamsianum* Rehd. & Wils.
R. *leishanicum* Fang & S. S. Chang,
  in prep.

UNPLACED SPECIES

R. *dimitrium* Balf. f. & Forr.
R. *nhatrangense* Dop in Lécomte
R. *potaninii* Batal
R. *purdomii* Rehd. & Wils.
R. *spilotum* Balf. f. & Forr.

EXCLUDED AND POORLY DESCRIBED
SPECIES

R. *blumei* Nutt.
R. *chlorops* Cowan
R. *dimidiatum* Balf. f.
R. *imberbe* Hutch.
R. *inopinum* Balf. f.
R. *kansuense* Millais
R. *magorianum* Balf. f.
R. *maximowiczianum* Lévl
R. *morseadianum* Millais
R. *paradoxum* Balf. f.
R. *peregrinum* Tagg
R. *planetum* Balf. f.
R. *pyrrhoanthum* Balf. f.
R. *serotinum* Hutch.
R. *venosum* Nutt.
R. *wallaceanum* Millais

Seedhead of R. *venator*

PART II SUBGENUS RHODODENDRON

The plants in this subgenus are the lepidote (with scales) rhododendrons. They are mainly the hardier rhododendrons, some originating from the coldest regions of the world, and they can vary from low-growing, often prostrate, shrublets just a few centimetres high to tree-like shrubs up to 9m (30ft) high. Their leaves are often aromatic, mostly evergreen and vary in size and form. The leaf buds are mainly convolute. The main characteristic of these species is their scales, which may be found on various parts of the plant. Some of the species have hairs as well as scales.

(Subgenus *Rhododendron* has three sections. Section *Rhododendron* has approximately 150 species, placed in 28 subsections. Section *Pogonanthum* has 13 species. These sections were revised by Dr. J. Cullen in his *Revision of Rhododendron I, Subgenus Rhododendron*. vol. 39. No. 1. (1980). The third section, *Vireya*, has approximately 200 species in 7 subsections and is currently being revised by Dr. Argent at the Royal Botanic Garden Edinburgh.)

SECTION RHODODENDRON

SUBSECTION AFGHANICA Cullen
Syn: Series *Triflorum*, Subseries *Hanceanum*

R. *afghanicum* Aitch. & Hemsl.

SUBSECTION BAILEYA Sleumer
Syn: Series *Lepidotum*, Subseries *Baileyi*

R. *baileyi* Balf. f.

SUBSECTION BOOTHIA Hutch. Sleumer
Syn: Series *Boothii*

R. *boothii* Nutt.
R. *chrysodoron* Tagg ex Hutch.
R. *dekatanum* Cowan
R. *leucaspis* Tagg
R. *megeratum* Balf. f. & Forr.
R. *micromeres* Tagg
R. *sulfureum* Franch.

SUBSECTION CAMELLIIFLORA (Hutch.) Sleumer
Syn: Series *Camelliaeflorum*

R. *camelliiflorum* Hooker f.

SUBSECTION CAMPYLOGYNA (Hutch.) Sleumer
Syn: Series *Campylogynum*

R. *campylogynum* Franch.

SUBSECTION CAROLINIANA (Hutch.) Sleumer
Syn: Series *Carolinianum*

R. *carolinianum* Hutch.
R. *minus* Michx.
   var. *minus*
   var. *chapmanii* (A. Gray) Duncan & Pullen

SUBSECTION CINNABARINA (Hutch.) Sleumer
Syn: *Cinnabarinum*

R. *cinnabarinum* Hooker f.
   ssp. *cinnabarinum*
   ssp. *xanthocodon* (Hutch.) Cullen
   ssp. *tamaense* (Davidian) Cullen
R. *keysii* Nutt.

SUBSECTION EDGEWORTHIA (Hutch.) Sleumer
Syn: Series *Edgeworthii*

R. *edgeworthii* Hooker f.
R. *pendulum* Hooker f.
R. *seinghkuense* Ward

SUBSECTION FRAGARIIFLORA Cullen
Part *Saluenense* Series

R. *fragariiflorum* Ward

SUBSECTION GENESTIERIANA (Hutch.) Sleumer
Syn: Series *Glaucophyllum*

R. *genestierianum* Forr.

SUBSECTION GLAUCA Sleumer
Syn: Series *Glaucophyllum*

R. *brachyanthum* Franch.
   ssp. *brachyanthum*
   ssp. *hypolepidotum* (Franch.) Cullen
R. *charitopes* Balf. f. & Farr.
   ssp. *charitopes*
   ssp. *tsangpoense*
R. *glaucophyllum* Rehd.
   var. *glaucophyllum*
   var. *tubiforme* Cowan & Davidian
R. *luteiflorum* Davidian
R. *pruniflorum* Hutch. & Ward
R. *shweliense* Balf. f. & Farr.

SUBSECTION HELIOLEPIDA (Hutch.) Sleumer
Syn: Series *Heliolepis*

R. *bracteatum* Rehd. & Wils.
R. *heliolepis* Franch.
   var. *heliolepis*
   var. *brevistylum* (Franch.) Cullen
R. *invictum* Balf. f. & Farr.
R. *rubiginosum* Balf. f. & Forr.

SUBSECTION LAPPONICA (Hutch.) Sleumer
Syn: Series *Lapponicum*

R. *amundsenianum* Hand.-Mazz.
R. *bulu* Hutch.
R. *burjaticum* Malyschev
R. *capitatum* Maxim.
R. *complexum* Balf. f. & W. W. Sm.
R. *cuneatum* Sm.
R. *dasypetalum* Balf. f. & Forr.
R. *fastigiatum* Franch.
R. *flavidum* Franch.
   var. *flavidum*
   var. *psilostylum* Rehd. & Wils.
R. *hippophaeoides* Balf. f. & W. W. Sm.
   var. *hippophaeoides*
   var. *occidentale* Philipson & Philipson
R. *impeditum* Balf. f. & W. W. Sm.
R. *intricatum* Franch.
R. *lapponicum* (L.) Wahl.
R. *minyaense* Philipson & Philipson
R. *nitidulum* Rehd. & Wils.
   var. *nitidulum*
   var. *omeiense* Philipson & Philipson
R. *nivale*
   ssp. *nivale*
   ssp. *boreale* Philipson & Philipson
   ssp. *australe* Philipson & Philipson
R. *orthocladum* Balf. f. & Forr.
   var. *orthocladum*
   var. *longistylum* Philipson & Philipson
   var. *microleucum* (Hutch.) Philipson & Philipson
R. *polycladum* Franch.
R. *rupicola* W. W. Sm.
   var. *rupicola*
   var. *chryseum* (Balf. f. & Ward) Philipson & Philipson
   var. *muliense* (Balf. f. & Forr.) Philipson & Philipson
R. *russatum* Balf. f. & Forr.
R. *setosum* D. Don
R. *tapetiforme* Balf. f. & Ward
R. *telmateium* Balf. f. & W. W. Sm.
R. *thymifolium* Maxim.
R. *tsaii* Fang
R. *websteranum*
   var. *websteranum* Rehd. & Wils.
   var. *yulongense* Philipson & Philipson
R. *yungningense* Balf. f. ex. Hutch.

SUBSECTION LEDUM (Genus *Ledum*)

R. *colombianum* (Piper) Harmaja
R. *groenlandicum* (Oeder) Kron & Judd
R. *hypoleucum* (Kron) Harmaja
R. *neoglandulosum* Harmaja (L. *glandulosum*)
R. *subarcticum* Harmaja (L. *decumbens*)
R. *tolmachevii* Harmaja (L. *macrophyllum*)
R. *tomentosum* (L. *palustre*) (L.)

SUBSECTION LEPIDOTA (Hutch.) Sleumer
Syn: Series *Lepidotum*

R. *cowanianum* Davidian
R. *lepidotum* Wall. ex G. Don
R. *lowndesii* Davidian

SUBSECTION MADDENIA (Hutch.) Sleumer
Syn: Series *Maddenii*

R. *amandum* Cowan
R. *burmanicum* Hutch.
R. *carneum* Hutch.
R. *ciliatum* Hooker f.
R. *ciliicalyx* Franch.
R. *ciliipes* Hutch.
R. *crenulatum* Hutch. ex Sleumer
R. *cuffeanum* Craib ex Hutch
R. *dalhousiae*
   var. *dalhousiae* Hooker f.
   var. *rhabdotum* (Balf. f. & Cooper) Cullen
R. *dendricola* Hutch.
R. *excellens* Hemsl. & Wils.
R. *fletcheranum* Davidian
R. *fleuryi* Dop
R. *formosum* Wall.
   var. *formosum*
   var. *inaequale* (Hutch.) Cullen
R. *horlickianum* Davidian
R. *johnstoneanum* (Watt ex) Hutch.
R. *kiangsiense* Fang
R. *levinei* Merr.
R. *liliiflorum* Lévl.
R. *lindleyi* Moore
R. *ludwigianum* Hosseus
R. *lyi* Lévl.
R. *maddenii*
   ssp. *maddenii* Hooker f.
   ssp. *crassum* (Franch.) Cullen
R. *megacalyx* Balf. f. & Ward
R. *nuttallii* Booth
R. *pachypodum* Balf. f. & W. W. Sm.
R. *pseudociliipes* Cullen
R. *roseatum* Hutch.
R. *rufosquamosum* Hutch.
R. *scopulorum* Hutch.
R. *surasianum* Balf. f. & Craib
R. *taggianum* Hutch.
R. *valentinianum* Forr. ex Hutch.
R. *veitchianum* Hooker
R. *walongense* Ward
R. *yungchangense* Cullen

SUBSECTION MICRANTHA (Hutch.) Sleumer
Syn. Series *Micranthum*
R. *micranthum* Turcz.

SUBSECTION MONANTHA Cullen
Syn: Series *Uniflora*

R. *concinnoides* Ward
R. *flavantherum* Ward
R. *kasoense* Ward
R. *monanthum* Balf. f. & W. W. Sm.

SUBSECTION MOUPINENSIA (Hutch.) Sleumer
Syn: Series *Moupinense*

R. *dendrocharis* Franch.
R. *moupinense* Franch.
R. *petrocharis* Diels

SUBSECTION RHODODENDRON
Syn: Series *Ferrugineum*

R. *ferrugineum* (L.)
R. *hirsutum* (L.)
R. *myrtifolium* Schott & Kotschy

SUBSECTION RHODORASTRA (Maxim.) Cullen
Syn: Series *Dauricum*
R. *dauricum* (L.)
R. *mucronulatum* Turcz.

SUBSECTION SALUENENSIA (Hutch.) Sleumer
Syn: Series *Saluenense*

R. *calostrotum* Balf. f. & Ward
   ssp. *calostrotum*
   ssp. *riparium* (Ward) Cullen
   ssp. *riparioides* Cullen
   ssp. *keleticum* (Balf. f. & Forr.) Cullen

R. saluenense Franch.
  ssp. *saluenense*
  ssp. *chameunum* (Balf. f. & Forr.)
  Cullen

SUBSECTION **SCABRIFOLIA**
(Hutch.) Sleumer
Syn: Series *Scabrifolium*

R. hemitrichotum Balf. f. & Forr.
R. mollicomum Balf. f. & W. W. Sm.
R. pubescens Balf. f. Forr.
R. racemosum Franch.
R. scabrifolium Franch.
  var. *scabrifolium*
  var. *spiciferum* (Franch.) Cullen
  var. *pauciflorum* Franch.
R. spinuliferum Franch.

SUBSECTION **TEPHROPEPLA**
(Hutch.) Sleumer
Syn: Series *Boothii*, Subseries
Tephropeplum

R. auritum Tagg
R. hanceanum Hemsl.
R. longistylum Rehd. & Wils.
R. tephropeplum Balf. f. & Forr.
R. xanthostephanum Merr.

SUBSECTION **TRICHOCLADA**
(Hutch.) Cullen
Syn: Series *Trichocladum* Balfour

R. caesium Hutch.
R. mekongense Franch.
  var. *mekongense*
  var. *melinanthum* (Balf. f. & Ward)
  Cullen
  var. *rubrolineatum* (Balf. f. & Forr)
  Cullen
  var. *longipilosum* (Cowan) Cullen
R. lepidostylum Franch.
R. trichocladum Franch.

SUBSECTION **TRIFLORA** (Hutch.)
Sleumer
Syn: Series *Triflorum*

R. ambiguum Hemsl.
R. amesiae Rehd. & Wils.
R. augustinii Hemsl.
  ssp. *augustinii*
  ssp. *chasmanthum* (Diels) Cullen
  ssp. *rubrum* (Davidian) Cullen
  ssp. *hardyi* (Davidian) Cullen
R. concinnum Hemsl.
R. davidsonianum Rehd. & Wils.
R. gemmiferum Philipson &
  Philipson
R. keiskei Miq.
R. lutescens Franch.
R. oreotrephes W. W. Sm.
R. pleistanthum Balf. f. ex Wilding
R. polylepis Franch.
R. rigidum Franch.
R. searsiae Rehd. & Wils.
R. siderophyllum Franch.
R. tatsiense Franch.
R. trichanthum Rehd.
R. triflorum Hooker f.
  var. *triflorum*
  var. *bauhiniiflorum* (Watt ex
  Hutch.) Cullen
R. yunnanense Franch.
R. zaleucum Balf. f. & W. W. Sm.

SUBSECTION **UNIFLORA** (Hutch.)
Sleumer
Syn: Series *Uniflorum*

R. ludlowii Cowan
R. pemakoense Ward
R. pumilum Hooker f.
R. uniflorum Ward
  var. *uniflorum*
  var. *imperator* (Hutch. & Ward)
  Cullen

SUBSECTION **VIRGATA** (Hutch.)
Cullen
Syn: Series *Virgatum*

R. virgatum Hooker f.
  ssp. *virgatum*
  ssp. *oleifolium* (Franch.) Cullen

SECTION **POGONANTHUM** D.
Don
Syn: Series *Anthopogon*

R. anthopogon D. Don
  ssp. *anthopogon*
  ssp. *hypenanthum* (Balf. f.) Cullen
R. anthopogonoides Maxim.
R. cephalanthum Franch.
  ssp. *cephalanthum*
  ssp. *platyphyllum* (Franch. ex Balf. f.
  & W. W. Sm.) Cullen
R. collettianum Aitch. & Hemsl.
R. fragrans (Adams) Maxim.
R. kongboense Hutch.
R. laudandum Cowan
  var. *laudandum*
  var. *temoense* Ward ex Cowan &
  Davidian
R. pogonophyllum Cowan & Davidian
R. primuliflorum Bur. & Franch.
R. radendum Fang
R. rufescens Franch.
R. sargentianum Rehd. & Wils.
R. trichostomum Franch.

SECTION **VIREYA**
Syn: Series *Pseudovireya*
R. christii Foerster

SUBGENUS **THERORHODION**
Syn: Series *Camtschaticum*

R. camtschaticum Pall (L.) (Small)
R. glandulosum (Standley ex Small) Q.
R. redowskianum (Maxim.) Hutch. Q.

PART III
AZALEA-TYPE
RHODODENDRONS

SUBGENUS **AZALEASTRUM**
SECTION **AZALEASTRUM**
Syn: Series *Ovatum*

R. bachii Lévl. Q.
R. hongkongense Hutch.
R. leptothrium Balf. f. & Forr.
R. ovatum Planch.
R. vialii Delavay & Franch.

SUBGENUS **AZALEASTRUM**
SECTION **CHONIASTRUM**
Syn: Series *Stamineum*

R. championiae Hooker
R. ellipticum Maxim.
R. esquirolii Lévl. Q.
R. feddii Lévl. Q.
R. hancockii Hemsl. Q.
R. henryi Hance Q.

R. latoucheae Franch. Q.
R. moulmainense Hooker
R. oxyphyllum Franch.
R. pectinatum Hutch. Q.
R. stamineum Franch.
R. stenaulum Balf. f. & W. W. Sm.
R. tainse Hutch. Q.
R. tutcherae Hemsl. & Wils. Q.
R. westlandii Hemsl.
R. wilsoniae Hemsl. & Wils.

SUBGENUS **CANDIDASTRUM**
Syn: Series *Albiflorum* Hooker
R. albiflorum Hooker

SUBGENUS **MUMEAZALEA**
Syn: Series *Semibarbatum*
R. semibarbatum Maxim.

Subgenus **PENTANTHERA**
Section **PENTANTHERA**
Syn: Series *Azalea*, Subseries *Luteum*

R. alabamense Rehd.
R. arborescens (Pursh) Torr.
R. atlanticum (Ashe) Rehd.
R. austrinum Rehd.
R. calendulaceum (Mich.) Torr.
R. canescens (Michx.) Sweet
R. cumberlandense
R. flammeum (Michx.) Sarg.
R. luteum Sweet
R. molle (Blume) G. Don
  ssp. *R. molle* (Blume) G. Don
  ssp. *R. japonicum* (A. Gray)
  Sutinger ex Wils.
R. occidentale A. Gray
R. periclymenoides (Michx.) Shinners
R. prinophyllum (Small) Millais
R. prunifolium (Small) Millais
R. viscosum (L.) Torr.

SUBGENUS **PENTANTHERA**
SECTION **RHODORA**
Syn: Series *Azalea*, Subseries *Canadense*

R. vaseyi A. Gray
R. canadense (L.) Torr.

SUBGENUS **PENTANTHERA**
SECTION **SCIADORHODIAN**
Syn: Series *Azalea*, Subseries *Schlippenbachii*

R. schlippenbachii Maxim.
R. albrechtii Maxim.
R. pentaphyllum Maxim.
R. quinquefolium Bisset & Moore

SUBGENUS **PENTANTHERA**
SECTION **VISCIDULA**
Syn: Series *Azalea*, Subseries *Nipponicum*

R. nipponicum Matsum

SUBGENUS **TSUTSUSI**
SECTION **BRACHYCALYX**
Syn: Series *Azalea*, Subseries *Schlippenbachii*

R. amagianum Makino
R. daiyunicum P. X. Tan
R. decandrum (Makino) Makino
R. dilatatum Miq.
R. farrerae Tate

R. hidakanum Hara
R. kiyosumense (Makino) Makino
R. mariesii Hemsl. & Wils.
R. mayebarae Nakai & Hara
R. nudipes Nakai
R. reticulatum D. Don ex G. Don
R. sanctum Nakai
R. viscistylum Nakai
R. wadanum Makino
R. weyrichii Maxim.

SUBGENUS **TSUTSUSI**
SECTION **TSUTSUSI**
Syn: Series *Azalea*, Subseries
Obtusum

R. apricum P. X. Tan
R. arunachalense Chambln. & Rae
R. atrovirense Franch. Q.
R. bicorniculatum P.X. Tan
R. boninense Nakai Q.
R. breviperulatum Hayata
R. chrysocalyx Lévl. & Van
R. chunii Fang Q.
R. cretaceum P.X. Tan
R. eriocarpum Hayata
R. florulentum P. X. Tan
R. flumineum Fang & M. Y. He.
R. fuchsiifolium Lévl.
R. gratiosum P. X. Tan
R. hainanense Merr., Philipp.
R. huiyangense Fang & M. Y. He
R. hunanense (Chun ex) P. X. Tan
R. indicum (L.) Sweet
R. jingpingense Fang & M. Y. He
R. jinxiuense Fang & M. Y. He
R. kaempferi Planch
R. kanehirae Wils.
R. kiusianum Makino
R. kwangtungense Merr. & Chun Q.
R. lasiostylum Hayata Q.
R. longiperulatum Hayata Q.
R. loniceriflorum P. X. Tan
R. macrosepalum Maxim.
R. mariae Hance Q.
R. meridionale P. X. Tan
R. microphyton Franch.
R. minutiflorum Hu Q.
R. mucronatum
R. myrsinifolium (Ching ex)
  Fang & He
R. naamkwanense Merr. Q.
R. nakaharai Hayata
R. noriakianum
R. oldhamii Maxim.
R. pulchroides Chung & Fang
R. rhuyuenense Chun in P. X. Tan.
R. rivulare Hand.-Mazz. Q.
R. rubropilosum Hayata
R. rufohirtum Hand.-Mazz. Q.
R. rufulum P. X. Tan
R. saxicolum Sleumer
R. scabrum G. Don
R. seniavinii Maxim. Q.
R. serpyllifolium Miq.
R. sikayotaizanense Masamune
R. simsii Planch
R. subcerinum P. X. Tan
R. subflumineum P. X. Tan
R. subenerve P. X. Tan
R. subsessile Rendle
R. taiwanalpineum
R. tashiroi Maxim.
R. tenuilaminare P. X. Tan
R. tosaense Makino
R. tschonoskii Maxim.
R. tsoi Merr.
R. tsusiophyllum Sugimoto
R. unciferum P. X. Tan
R. viscigemmatum P. X. Tan
R. yaoshanicum Fang & M. Y. He
R. yangmingshanense P. X. Tan
R. yedoense var. *poukhanense* Maxim.

GLOSSARY

anthocyanins water-soluble, red- to blue-coloured plant colouring matter

basal of/at the base

branchlet small branch, off a larger branch

bullate blistered or puckered

calyx outer whorl of a flower

campanulate shaped like a bell

carotinoids pigments including carotene (orange or red substance)

carpel one of the parts of the flower's compound ovary

chloroplast small body in cytoplasm of plant cell containing chlorophyll (colouring matter of green parts of plant)

chromoplasts yellow/orange organelles in plant cells, normally developing out of chloroplasts after the breakdown of chlorophyll (after which only yellow/orange carotinoids have a colouring function)

coriaceous leathery

corolla whorl of petals, forming inner envelopes of plant

crenate with notched edge

dendroid tree-shaped

elepidote without scales

elliptic leafshape: with curved sides tapering towards both ends, widest part in the middle; about twice as long as wide

entire undivided (margin)

epiphyte a plant growing on another plant (not a parasite)

glabrous without hairs

glaucous with greyish, waxy bloom

indumentum a woolly or hairy covering, particularly of stems and leaves

inflorescence group of flowers, flowers from one bud

lacerate torn or irregularly cleft (of scale)

lanceolate leafshape: lance-shaped; the leaf is widest below the middle, and about three times as long as wide

lepidote scaly, with minute scales

linear leafshape: a pencil-like leaf, narrow with parallel sides, tapering at both ends; about 10–12 times as long as wide

lobe flat or round projecting or pendulous part

oblanceolate leafshape: the leaf is widest above the middle, with a broad apex, and tapering towards the stalk; about three times as long as wide

oblong leafshape: sides of leaf are approximately parallel, tapering towards both ends which are rounded or obtuse; leaf is about twice as long as wide

R. trichanthum Rehd. "with hairy flowers"

obovate leafshape: the leaf is wider above the middle, with broad apex, and tapering towards the stalk; like oblanceolate, but only twice as long as wide

orbicular leafshape: round

ovate leafshape: the sides are rounded, tapering towards apex, widest part near the base; longer than wide

oval leafshape: with curved sides, rounded at both ends, middle is widest; longer than wide

pedicel stalk of flower

petiole stalk of leaf

procumbent growing along ground

propagule part of plant, such as bud, that may separate from the parent plant and grow on by itself

pubescent with short soft hairs

rachis main axis of an inflorescence

sessile stalkless

spathulate with a broadly rounded apex gradually tapering towards the stalk

stipitate gland gland of stipe (stalk or stem supporting carpel)

stoloniferous sending out runners from the stem of the plant, often underground

style thread-like part of a pistil between ovary and stigma

terminal at the end (of a shoot)

tomentum a dense covering of hairs

truss a single inflorescence

vesicular of vesicle (small or hollow structure)

whorl ring of leaves around stem of plant

BIBLIOGRAPHY

Bean, W. J., *Trees and Shrubs in the British Isles,* Vol. 3, 8th edition, London: John Murray, 1976

Berg, Johann & Lothar Heft, *Rhododendron und immergrüne Laubgehölze* (Rhododendrons and evergreen broad-leaved trees); third edition, Stuttgart: Eugen Ulmer GmbH 1991

Chamberlain, D. F., *Notes from the Royal Botanic Garden Edinburgh,* Vol. 39 No. 2, Edinburgh 1982, "A revision of *Rhododendron* II Subgenus *Hymenanthes*"

Chamberlain, D. F. & S. J. Rae, *Edinburgh Journal of Botany,* Vol. 47, No. 2, Edinburgh 1990, "A revision of *Rhododendron.* IV. Subgenus *Tsutsusi*"

Cox, Euan H. M. and Peter A. Cox, *Modern Rhododendrons,* Edinburgh: Thomas Nelson & Sons Ltd., 1956

Cox, Peter A., *Dwarf Rhododendrons,* London: Batsford, 1973

Cox, Peter A., *The Encyclopedia of Rhododendron Hybrids,* London: Batsford, 1988

Cox, Peter A., *The Larger Rhododendron Species,* London: Batsford, 1988

Cox, Peter A., *The Smaller Rhododendrons,* London: Batsford, 1988

Cox, Peter & Kenneth Cox, *Cox's Guide to Choosing Rhododendrons,* London: Batsford, 1990

Cox, Peter A., *The Cultivation of Rhododendrons,* London: Batsford, 1993

Cullen, J., *Notes from the Royal Botanic Garden Edinburgh,* Vol. 39. No. 1, Edinburgh 1980, 'A revision of *Rhododendron.* I. Subgenus *Rhododendron,* sections *Rhododendron* and *Pogonanthum*'

Davidian, H. H., *The Rhododendron Species,* Vols I–III, London: Batsford, 1983, 1989, 1992

Fletcher, H. R. *A Quest of Flowers,* Edinburgh: Edinburgh University Press, 1995

Greer, Harold E., *Greer's Guidebook to Available Rhododendron Species & Hybrids,* Eugene, Oregon: Offshoot Publications, 1980; third edition

Greer, Harold E. & H. Salley, *Rhododendron Hybrids,* London: Batsford, 1992

Lancaster, C. Roy, *Plant Hunting in Nepal,* London: Croom Helm, 1981

Lancaster, C. Roy, *Travels in China,* Woodbridge: Antique Collector's Club, 1989

Leach, David G., *Rhododendrons of the World,* London: Allen & Unwin, 1962

Michaux, André, *Flora Boreali Americana,* 2 volumes, France: Apud Fratres Levrault, anno xi, 1803

Millais, J. G., *Rhododendron Species and the Various Hybrids,* Vol. 1, London: Longmans, 1917; Vol. 2: London: Green, 1924

Phillips, Lucas, & Peter N. Barber, *The Rothschild Rhododendrons,* London: Cassell, 1967

Royal Horticultural Society, London, *Rhododendron Yearbook* 1946–1971; *Rhododendrons* 1972–1973; *Rhododendrons with Magnolias and Camellias,* 1974 ff.

Schneider, Camillo Karl, *Illustriertes Handbuch der Laubhölzkunde* (Illustrated Handbook of the Study of Deciduous Trees), Jena: Verlag Gustav Fischer, 1912

Spethmann, Wolfgang, Manfred Oetting & Burkhard Walter, *Rhododendron Bibliography,* published by the Institute for Fruit and Nursery Science, University of Hanover, and the German Rhododendron Society, Bremen; 270pp ring-bound; available from the Institute for Fruit and Nursery Science, University of Hanover, Am Steinberg 3, D 31157 Sarstedt. Tel. +49 (0) 5066 – 82 61 14; Fax +49 (0) 5066 – 4261.

Stearn, William T., *Botanical Latin,* Newton Abbot: David & Charles, fourth edition 1992

PERIODICALS

American Rhododendron Society Quarterly Journal, Vol. 36, No. 2, Spring 1982 for "Plant hunting in China" by Peter A. Cox

The Quarterly Bulletin of the American Rhododendron Society, Vol. 2, No. 4, Nov. 20, 1948, p69 in *Quarterly Reprint of the American Rhododendron Society Bulletin* (three reprinted volumes: 1947, 1948 and 1949) for "Dr. Joseph E. Rock letter"

The Rhododendron and Camellia Yearbook 1971, London: The Royal Horticultural Society, 1970

The Rhododendron Society Notes, Vol. II, Part III, 1922, Ipswich: W. S. Cowell Ltd, 1922

Rhodora 14:99, 1912, Cambridge, Massachusetts: the New England Botanical Club for Alfred Rehder's article

INDEX

GUIDE TO SIZE OF PAINTINGS

55 × 38 centimetres (22 × 15 inches)
R. arboreum and *varieties, R. auriculatum, R. campanulatum, R. crassum, R. decorum, R. edgeworthii, R. facetum, R. falconeri,* Foliage Studies, *R. fulvum, R. griersonianum, R. insigne, R. irrorata, R. lanatum, R. lindleyi, R. ponticum, R. sinogrande, R. souliei, R. thomsonii, R. westlandii, R. yakushimanum.*

39 × 28 centimetres (15½ × 11 inches)
R. baileyi & *R. sulfureum, R. barbatum, R. bureavii, R. camelliiflorum, R. campylogynum, R. camtschaticum, R. christii, R. cinnabarinum, R. desquamatum, R. ferrugineum, R. fulgens, R. glaucophyllum* etc, *R. glishrum, R. lapponicum* etc, *R. lepidostylum, R. luteum, R. micranthum, R. minus, R. moupinense, R. mucronulatum, R. neriiflorum, R. pemakoense, R. pseudochrysanthum, R. reticulatum, R. saluenense, R. scabrifolium* etc, *R. schlippenbachii, R. selense, R. semibarbatum, R. tephropeplum, R. vaseyi, R. venator, R. williamsianum, R. yunnanense.*

ACKNOWLEDGEMENTS

I would like to thank all who made this book a reality, especially Mr. and Mrs. Edmund de Rothschild and family for the freedom of their beautiful Garden at Exbury; Mr. Nicholas de Rothschild for the loan of the paintings on pages 43, 45 and 59; Mr. Kenneth Lowes for his careful guidance; Mr. Douglas Harris and Garth and Deirdre Vaughan-Brown for their encouragement; Mr. Peter Cox, Glendoick Garden, Mr. Keith Rushforth, Mr. Ivor Stokes, Swansea Botanical Complex, Mr. Mark Sparrow, Royal Botanic Gardens Kew, Mr. Mark Flanagan, Royal Botanic Gardens, Wakehurst for supplying speciments and materials from their collections, Mr. Ian McLachlan and all the staff at Exbury Gardens; Mrs. Christine Dykes, Pamela, Kristen and Andrew for collecting species further afield; Mrs. Sylvia

Goulding (Silva Editions); Anna Mumford, Jo Weeks and Sarah Widdicombe of David & Charles; Mrs. Rosemary Seys, whose friendship helped at a most critical time.
I give my sincere gratitude for the help, guidance and specimens, given so generously by The Royal Botanic Gardens, Edinburgh, especially Dr. D. Chamberlain, Mr. Roger Hyam and a special thank you to Mr. Colin Belton who collected and packed the species so carefully for their journey south.
Finally, I give my deepest and most sincere thanks, to that happy band of rhododendron species enthusiasts, my contributors, whose lovely articles, encouragement and assistance made this book a pleasure to create.